T0092064

Neurorobotics

Intelligent Robotics and Autonomous Agents
Edited by Ronald C. Arkin

A complete list of the books in the Intelligent Robotics and Autonomous Agents series appears at the back of this book.

Neurorobotics

Connecting the Brain, Body, and Environment

Tiffany J. Hwu and Jeffrey L. Krichmar

The MIT Press
Cambridge, Massachusetts
London, England

The MIT Press would like to thank the anonymous peer reviewers who provided comments on drafts of this book. The generous work of academic experts is essential for establishing the authority and quality of our publications. We acknowledge with gratitude the contributions of these otherwise uncredited readers.

This book was set in Times New Roman by Westchester Publishing Services. Printed and bound in the United States of America.

Library of Congress Cataloging-in-Publication Data

Names: Hwu, Tiffany, author. | Krichmar, Jeffrey L. author.
Title: Neurorobotics : connecting the brain, body, and environment / Tiffany Hwu and Jeff Krichmar.
Other titles: Intelligent robotics and autonomous agents
Description: Cambridge, Massachusetts : The MIT Press, [2022] | Series: Intelligent robotics and autonomous agents | Includes bibliographical references and index.
Identifiers: LCCN 2021060534 (print) | LCCN 2021060535 (ebook) | ISBN 9780262047067 (hardcover) | ISBN 9780262370530 (epub) | ISBN 9780262370547 (pdf)
Subjects: MESH: Neurosciences—methods | Robotics—methods | Machine Learning | Equipment Design
Classification: LCC QP360.7 (print) | LCC QP360.7 (ebook) | NLM WL 26.5 | DDC 612.80285—dc23 /eng/20220131
LC record available at https://lccn.loc.gov/2021060534
LC ebook record available at https://lccn.loc.gov/2021060535

10 9 8 7 6 5 4 3 2 1

Contents

Preface

Neurorobotics is an interdisciplinary field that touches on artificial intelligence, cognitive science, computer science, engineering, psychology, neuroscience, and robotics. Neurorobots are autonomous systems modeled after aspects of the brain. Because the brain is so closely coupled to the body and the body is situated in the environment, these robots can be a powerful tool for studying neural function in a holistic fashion. Neurorobots may also be a means to develop autonomous systems that have intelligence rivaling that of biological organisms. This book broadly covers the field of neurorobotics and the methods that neuroroboticists employ. Concepts regarding algorithms, experimental design, embodiment, robot construction, and computer programming are discussed. The book describes approaches and design principles grounded in biology and neuroscience to develop intelligent autonomous systems.

In the first part of the book we cover the background knowledge from these diverse fields that is needed to conduct neurorobotic experiments or just to understand the field of neurorobotics. In the second part we discuss several principles that should be considered in designing neurorobots and experiments using robots to test brain theories. The chapters in this part are strongly inspired by Pfeifer and Bongard's (2006) design principles for intelligent agents. We build on these design principles by grounding them in neuroscience and by adding principles based on neuroscience research. In the third part we introduce examples of neurorobots for navigation, developmental robotics, and social robots. Throughout the book, we provide the cognitive science and neuroscience background that inspires these robots. We end the book with a summary and speculation of where we believe neurorobotics can make major practical contributions.

In addition to these chapters, we provide videos and neurorobot simulations on our supplementary materials website https://faculty.sites.uci.edu/krichmarlab/. The videos were created for lectures to our undergraduate and graduate Cognitive Robotics class at the University of California, Irvine. The website also provides a link to software repositories that contain neurorobot examples. These examples are referenced in boxed text throughout the chapters. The examples, written in the Python and C programming languages, are designed to run in the Webots robot simulator (an open source platform maintained by Cyberbotics Ltd.). The book was largely written during the COVID-19 pandemic while courses were conducted virtually and remotely. We hope to add to these repositories after our robot classes have returned to in-person presentation with physical robots. We encourage others to contribute their examples to the website. If you are interested, you are welcome to contact the authors.

The intended audience for this book includes everyone who has an interest in exploring neurorobotics for their research or just for their edification. We have used the material in this book along with supplemental programming and supplemental projects to teach cognitive robotics to undergraduate students and graduate students in cognitive science, computer science, psychology, and other disciplines.

We hope you enjoy this book as much as we enjoy this fun and exciting topic!

I NEURОROBOT BACKGROUND AND FOUNDATIONS

1 Neurorobotics: Origins and Background

1.1 Neurorobotics

Neurorobots* are robots whose control has been modeled after some aspect of the brain. Because the brain is so closely coupled to the body and the body is situated in the environment, neurorobots can be a powerful tool for studying neural function in a holistic fashion. Neurorobots may also be a means to develop autonomous systems that have some level of biological intelligence. Neurorobotics is a subdiscipline of the broader field of **cognitive robotics**. Cognitive robotics emphasizes biologically inspired behavior and intelligence (Cangelosi & Asada, 2022). It is an interdisciplinary approach with contributions from cognitive science, neuroscience, and social sciences. Neurorobotics, as well as other subfields such as developmental robotics, evolutionary robotics, soft robotics, and swarm robotics, to name a few, have a broad common goal to create biologically inspired robots. Although this book touches on some of these other approaches, it concentrates on the neuroscience underlying and inspiring neurorobots.

The motivation to study neurorobotics comes from a desire both to understand cognition and to improve autonomous applications. From the perspective of cognitive science and neuroscience, robots allow us to model the brain in an embodied environment. In technology there is increasing interest in moving robots out of factories and into homes and businesses. In interaction with people, safety standards become increasingly important and the environment becomes less predictable. To meet these needs, researchers aim to develop artificial general intelligence (AGI), the ability to apply common sense, make new inferences on the basis of previous experience, and communicate as humans can. Achieving AGI will rely heavily on developments in cognitive robotics and neurorobotics.

Neurorobotics focuses on the development of learning, reasoning, exploring, and communicating. Learning is an essential skill for a neurorobot, as it must adapt to changes in the environment and be able to reuse previously learned tasks in novel situations. Reasoning allows neurorobots to make inferences about the world and change their behaviors

* Bold-faced font denotes terms defined in the glossary.

accordingly. Exploring is how neurorobots learn about a new environment and act upon inferences. Finally, communication allows the neurorobots to interface with humans and other **agents**, enabling teamwork and cooperation.

This book explains how neurorobots can be used to study the brain and behavior through examples, which like the brain can be quite complex, and through experiments, which we have aimed to make simple. Throughout the book, we introduce specific neurorobot examples to describe some of the basic principles of neurorobotics. In the companion website, we provide representative examples of neurorobots that can be carried out by interested readers (see https://faculty.sites.uci.edu/krichmarlab).

With this book we aim to achieve several goals: (1) to introduce the field of neurorobotics; (2) to define and discuss the principles we believe are important for designing neurorobots; (3) to provide a unique understanding of how brain activity leads to interesting behavior; (4) to provide a method for designing intelligent robots; and (5) to inspire researchers, students, and hobbyists to use the neurorobotic approach for their projects.

1.2 Early Examples of Neurorobots

Many believe that neurorobotics got its start with Grey Walter's tortoises,[1] built in the late 1940s. These robots had rudimentary light sensors and collision detectors controlled by a simple analog circuit (figure 1.1). However, these simple brains produced seemingly complex behavior that we might call intelligent (Holland, 2003). The tortoises, called Elsie and Elmer, had photoelectric sensors that caused the steering mechanisms to move the tortoises toward

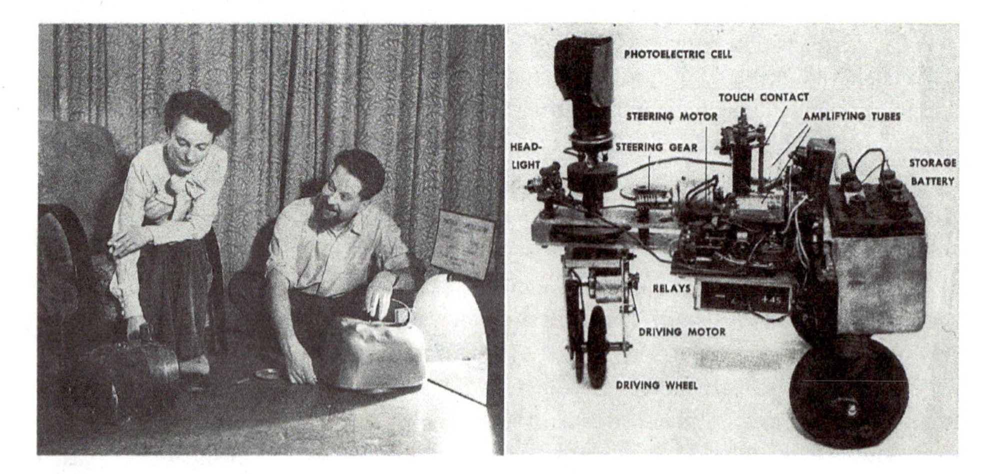

Figure 1.1
Grey Walter's tortoises. *Left*: Grey Walter and his wife Vivian with the Elsie and Elmer tortoises. *Right*: Labeled photo of Elsie without her shell. Photos taken from http://cyberneticzoo.com.

light sources. If the shell hit an obstacle, contact was made with a switch, causing the tortoise to back and turn away from the obstacle. When Elsie's batteries were low, she returned to her charging hutch, which was signaled by a light source. The important lesson from Elsie and Elmer was that interaction with the environment, even if controlled by a small, simple nervous system, can result in realistic-looking behavior.

Another seminal moment in the history of neurorobotics was the *Vehicles* thought experiments conceived by renowned neuroanatomist Valentino Braitenberg (Braitenberg, 1986). Each chapter of his short book introduced a simple robot or vehicle that demonstrated a lesson in neuroscience; for example, connecting the left light sensor to the right motor of the imaginary robots and vice versa, Braitenberg described the difference between contralateral and ipsilateral connections and their effects on behavior. Using vehicles, he introduced concepts of sensorimotor loops, inhibition, and valence with these simple thought experiments.

For example, vehicle 2 has two sensors and two motors on each side (see figure 1.2, left). For simplicity, assume that each motor is attached to a wheel. The left sensory neuron has a connection to the left motor neuron, which is an **excitatory connection** as denoted by the plus sign, and a similar sensory to motor connection is set for the right side. In neuroscience, connections that stay on the same side of the body are called **ipsilateral connections**. The more light that hits the sensory neuron, the more it excites the motor neuron and the faster the wheel moves. If the light source is closer to the right side of the

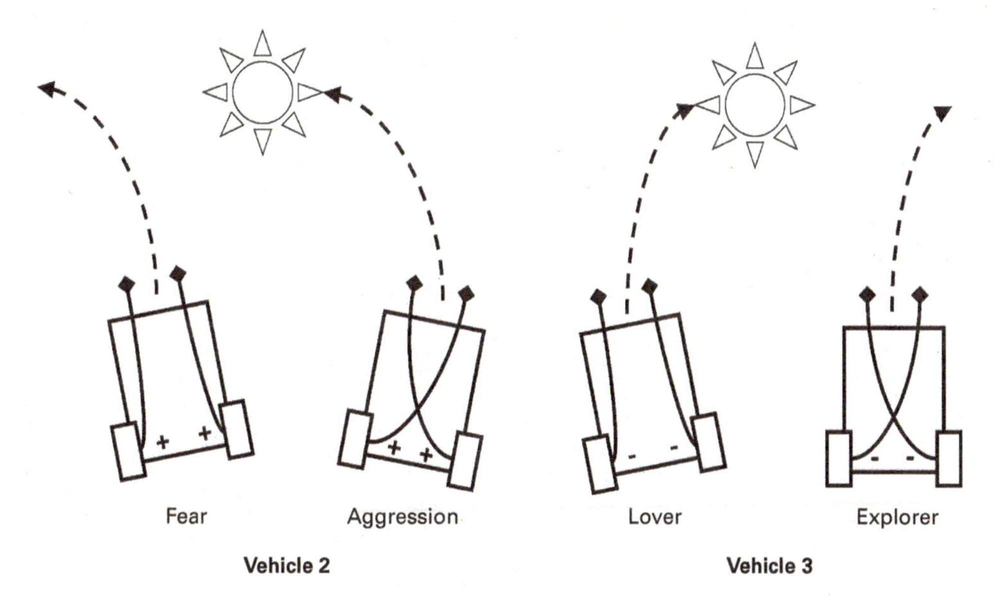

Figure 1.2
Braitenberg's vehicles 2 and 3. Each vehicle has two light sensors and two motors attached to wheels. Plus signs (+) denote excitatory connections, and minus signs (–) denote inhibitory connections.

vehicle, the right wheel turns faster than the left wheel, causing the vehicle to move away from the light source (i.e., fear). If we change the connections to be **contralateral connections**, in which the left sensory neuron is connected to the right motor neuron and vice versa, the behavior changes from moving away from the light source to moving toward the light source (i.e., aggression). Within this simple thought experiment is a valuable lesson about neuroanatomy. Most connections in the nervous system cross from one side to another; that is, they are contralateral. For example, the left side of our brain for motor and sensory information sends information to and gets information from the right side of the body. However, some connections stay on the same side, that is, they are ipsilateral. Each of these projections provides different signals about the environment in relation to one's body. This information can be used to produce a range of actions and responses, in this case, orienting or avoiding.

In vehicle 3, the connections are changed from excitatory to an **inhibitory connection** as denoted by a minus sign (see figure 1.2, right). Ipsilateral connections cause the vehicle to move toward the light source and to slow down as it gets closer to the light source (i.e., lover). Making the connections contralateral causes the vehicle to slow down as it approaches the light source and then move away (i.e., explorer). Throughout the brain and nervous system, there is a balance of excitatory and inhibitory connections that are crucial for generating, releasing, and suppressing behaviors.

Box 1.1 describes a Webots simulation of Braitenberg's vehicles that can be tried out by the interested reader.

These early examples follow the idea of neurorobotics in which artificial nervous systems, even simple ones, gives rise to lifelike behavior. For a historical view on neurorobotics, the interested reader is referred to a review by one of the authors (Krichmar, 2018).

Box 1.1

A Webots simulation of vehicles is provided in the Braitenberg vehicles folder on GitHub at https://github.com/jkrichma/NeurorobotExamples/.

Braitenberg wrote the book *Vehicles: Experiments in Synthetic Psychology* to provide neuroscience lessons with simple robots. In this simulation, we use Webots to create Braitenberg's vehicles 2 and 3 (see figure 1.3). We test how simulating the connectivity (i.e., contralateral or ipsilateral) and the connection type (i.e., excitatory or inhibitory) alters overall behavior. The robot is blue, and its wheels are the red circles. The yellow Xes and lines coming out of the robot denote the location of the light sensors. The light source is the yellow sun symbol.

Figure 1.3 shows sample runs of the vehicles simulation. The robot starts at an initial position away from the light source. In vehicle 2A, increasing excitation of the left light sensor causes the left wheel to move faster than the right. The robot moves away from the light source and to the right. In vehicle 2B, the left light sensor is connected to the right wheel and vice

Box 1.1
(continued)

Figure 1.3
Webots simulation of Braitenberg's vehicles. The blue robot either approaches or avoids the light source (sun symbol). The yellow rays emanating from the robot show the angle of the robot's light sensors.

versa. Now the robot moves toward the light source. In Vehicles 3A and 3B, the more the light sensor is excited, the slower the wheels move. Note how this changes the robot's behavior.

In the Webots simulation, try moving the sun around with the different vehicle types. See how the behavior differs. The provided Python code implements one possible way to code the vehicles. Can you think of other ways to create vehicles 2 and 3? Your solution should be as simple as possible, that is, the behavior should be captured in just a few lines of code.

1.3 Robotics

To elicit behavior, a system needs **sensors** to receive stimuli from the world and **actuators** to act in the world. Robots, in general, provide a surrogate body to perform neurorobotic experiments. Before delving further into neurorobotics, we first look at the field of robotics and the tools it provides. We give a brief overview of the different methods for controlling these robots with **artificial intelligence** and other approaches.

Robots are well established in many industries today. For instance, robots can perform routines with high precision, automating many processes in assembly lines for manufacturing. To achieve such performance, traditional methods in robotics have been developed to perform movements with very low error, achieving great results. However, when it comes to interacting with the environment in an intelligent manner, robots face numerous challenges given the uncertainty and dynamic nature of the world.

For the purposes of this book we use robots as a tool for running neurorobot experiments in which the behavior of the robot is examined as a behavioral neuroscientist or cognitive scientist would examine behavior when one of their subjects is challenged with problems or stimuli. However, to conduct neurorobotics experiments we need a basic understanding of robots and their capabilities.

1.3.1 Origins, Benefits, and Ethical Considerations

The word *robot* was used to describe mechanical workers in a Czechoslovakian science fiction play called *R.U.R.* or *Rossum's Universal Robots* by the Czech writer Karel Čapek that premiered in 1921. The Czech word *robota* means *forced labor*. In the play, androids called robots are created to work in a factory. However, they become unhappy with their lot in life, rise up in rebellion, and end the human race. Unfortunately, these dark tales of robots at odds with humans continue in today's entertainment industry with the *Terminator* series, *Ex Machina*, and *Blade Runner*, to name a few (see figure 1.4). Moreover, technology experts such as Elon Musk, Stephen Hawking, and Bill Gates warn against the potential dangers of robots and artificial intelligence.[2]

In his 1950 book *I, Robot*, the great science fiction writer Isaac Asimov defined three laws of robotics to ensure safety during interactions with robots (40):

1. A robot may not injure a human being or, through inaction, allow a human being to come to harm.

2. A robot must obey orders given to it by human beings except where such orders would conflict with the First Law.

3. A robot must protect its own existence as long as such protection does not conflict with the First or Second Law.

Although the laws seem well defined, there are gray areas in which conflicts can emerge. Asimov explored these conflicts in his stories as the humans interact with the robots. It

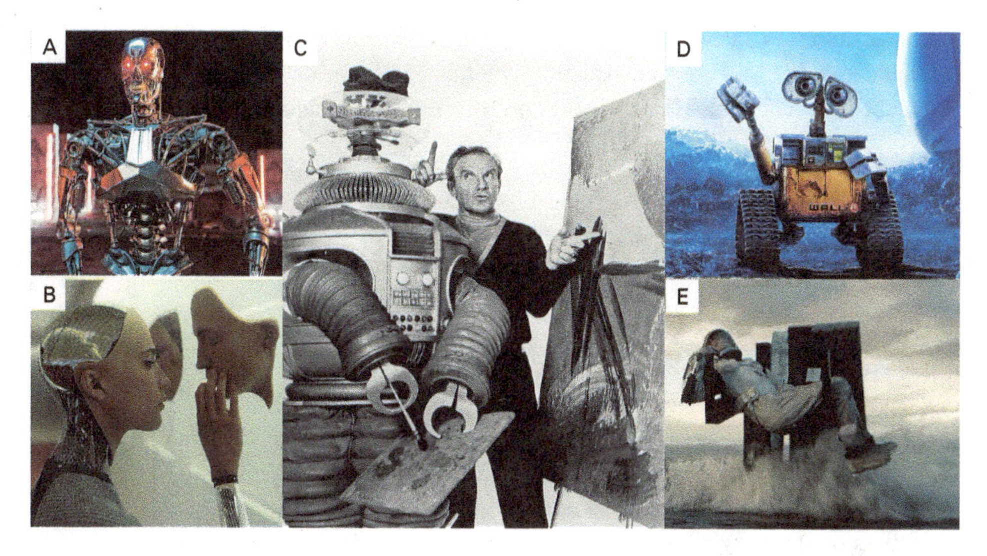

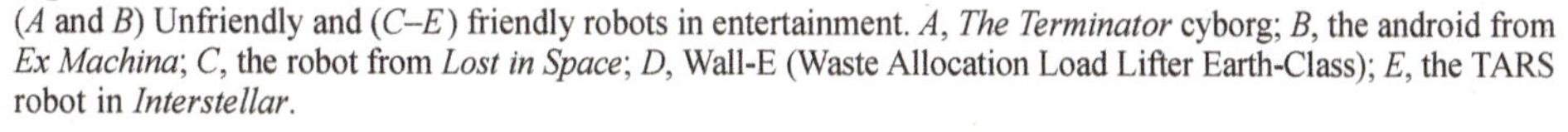
Figure 1.4
(*A* and *B*) Unfriendly and (*C–E*) friendly robots in entertainment. *A*, *The Terminator* cyborg; *B*, the android from *Ex Machina*; *C*, the robot from *Lost in Space*; *D*, Wall-E (Waste Allocation Load Lifter Earth-Class); *E*, the TARS robot in *Interstellar*.

should be noted in these books that the issue is usually not with the robots but with the position into which the humans put the robots. Still, the current development of automated weapons could potentially violate Asimov's first law. The robotics and AI communities raised awareness of these issues in dramatic form with their Slaughterbots scenario.[3] Although beyond the scope of this book, the topic of machine ethics is very interesting and is an active area of research. Interested readers may want to read the point of view by roboticist Ron Arkin (Arkin, 2016; Borenstein & Arkin, 2016).

Of course, we are of the opinion that the benefits of robots outweigh the risks (Markoff, 2015). We prefer the depiction of Wall-E and the helpful robot in *Interstellar* over the more negative images of robots and AI in entertainment (see figure 1.4). Robotics has potential benefits for healthcare, the elderly, manufacturing, and a wide range of high-risk jobs. For example, the tsunami that damaged the Fukushima Daiichi nuclear power plant in 2011 led to the robotics community further addressing how robots might be better equipped for assisting in disaster relief and rescue operations. During the COVID-19 pandemic, robots were used to clean hospitals and other common areas as well as to connect people who were locked down and could not see friends and family (Yang et al., 2020).

1.3.2 Autonomy

Because organisms with nervous systems are autonomous agents, we mainly consider autonomous robots in this book. For a more complete treatment of **autonomous robots**, we recommend a book on this topic by roboticist George Bekey (2017). Bekey defines a robot as a machine that senses, thinks, and acts. An autonomous system is defined as capable of operating in the real-world environment without any form of external control for an extended period of time. This certainly fits with biological organisms and is a major goal of neurorobotics.

1.3.2.1 Sensors and actuators

To operate in the real world, robots need a way to take in information from the environment through active sensing and then act in the world on the basis of these stimuli. For robotics, this requires sensors and actuators. As we briefly describe, these come in many forms and have similarities to those in biological organisms.

Sensors come in two forms, **exteroceptive** and **interoceptive** (see figure 1.5). Exteroceptive sensors obtain information from the external environment. In biology these include our senses of vision, hearing, smell, touch, and taste. The analogous sensors often used in robots include cameras (vision), microphones (hearing), and switches (touch). Less usual sensors include artificial skin (fine touch) and chemical sensors (smell). Nonbiological sensors such as lidar, a laser-enabled distance sensor, and global positioning systems (GPS) are often used in today's robots.

Interoceptive sensors monitor the internal system. In biology, neurons monitor our bodily functions and body position. An example is the inner ear or vestibular system, which is important for balance and gives us a signal of acceleration. The equivalent of a vestibular system in robots is an **inertial measurement unit** (IMU). An IMU typically packages an accelerometer, compass, and gyroscope to give a sense of internal force and direction. Other interoceptive sensors in biology include proprioception, which gives information of the forces and loads on our joints and muscles, and the autonomic nervous system, which monitors

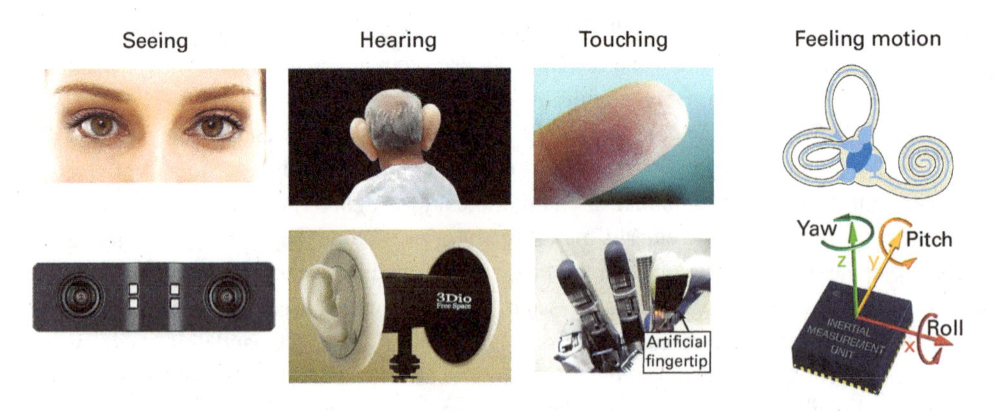

Figure 1.5
Sensors in biology (*top*) and for robotics (*bottom*). Seeing: Three-dimensional vision can be mimicked with stereo cameras (eYs3D stereo camera, Sparkfun Electronics). Hearing: Two ears for hearing and sound localization can be realized with binaural microphones (from 3Dio). Touching: Fine touch can be realized with artificial fingertips (adapted from Bologna et al. [2011]). Feeling motion: The inner ear or vestibular system, which provides a signal of self-motion, can be achieved with an inertial measurement unit.

energy levels, temperature, heart rate, and other functions. For robotics, there are rotation sensors and load sensors, which may be analogous to proprioception, as well as battery monitors, temperature gauges, and so on to monitor system health.

Actuators provide a means to physically interact with the environment. In most cases, this includes electric motors that power wheels, legs, or arms. Motors such as these have limitations in compliance, flexibility, and power. Therefore, alternative means for moving parts is an active area of research. Engineers have been developing artificial muscles by using pneumatic and hydraulic actuators. Research in electroactive polymers has the potential to create something close to a muscle. Still, actuators that have the range of motion, the compliance, size-to-power ratio, and variable load found in our muscles are currently out of reach. One promising direction for compliant actuators is the development of soft materials (Trimmer et al., 2015). This may provide the flexibility that is missing from many actuators. Furthermore, like our skin and muscle, sensors can be embedded in the actuator itself.

1.3.2.2 Control and software approaches

To be autonomous, the robot's behavior and responses must be guided by some control system. For a biological organism, this is its nervous system. Throughout the book, we introduce neurorobotic examples that create artificial nervous systems to control robots. In any system, there are levels of control that support autonomy (see figure 1.6). At the lowest level, there are monitors and drivers to send motor commands to the robot and receive sensory information from the robot. For example, if you want the robot's wheels to move at 50 percent speed for 2 seconds, there must be a command that when executed causes the robot to move its wheels appropriately. Similarly, if you want to know the light level in a room, there must be a command to read from the light sensor. In addition, you may want to know the battery level of your robot. At the intermediate level, one can build up functionality by using sensory readings to guide robot actions. For example, distance sensors may be read to notify the controller that there is an obstacle on one side, causing the robot to avoid the obstacle by turning to the other side; in another example, a camera may detect an interesting object causing the robot to approach that object for closer inspection. These intermediate functions can be combined to create full suites of behavior at the highest level of control. Decision making, long-term planning, and goal-driven behavior are a few of the high-level functions that an autonomous robot might demonstrate. Information flows from high to low, as well. A high-level planner may need to send commands to the intermediate level navigation system, which will plan routes and avoid obstacles. The route is a series of low-level motor commands to move toward intermediate goal locations, which may be interrupted by sensor readings identifying an obstacle, which in turn causes the route to be replanned.

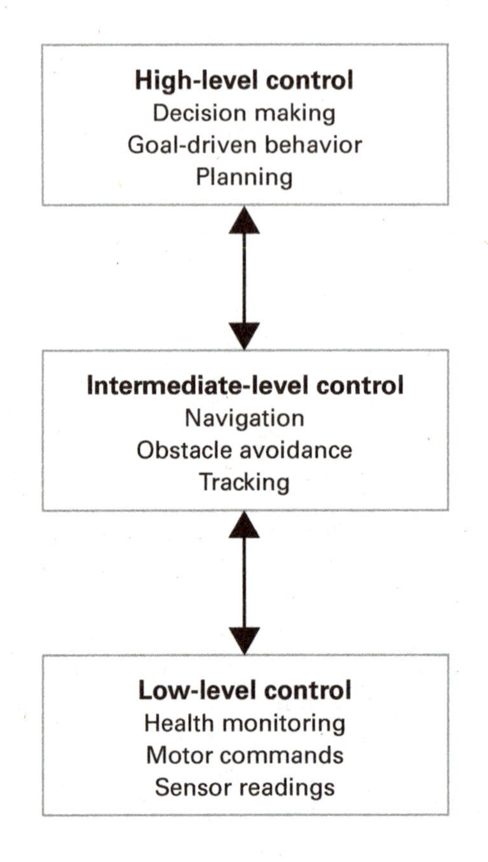

Figure 1.6
Control levels for autonomy. Redrawn from Bekey (2017).

1.3.2.3 Levels of control in neuroscience

These levels of control have parallels to the self-monitoring systems in biology (Chiba & Krichmar, 2020). In the nervous system there is low-level control for sensor processing and motor reflexes. At the intermediate level the autonomic nervous system maintains homeostasis and predicts system set points. At the highest, cognitive level, planning, and prediction can further monitor the system and optimize performance. It should be emphasized that there is information flow between these levels from both a systems neuroscience and an engineering point of view. Figure 1.7 provides an overview of self-monitoring in biology and engineering. The left side of figure 1.7 lists components of the nervous system involved in primitive reflexive behavior and sensory processing (bottom left), the maintenance of system stability (middle left), and higher-level planning and control (top left). The right side of figure 1.7 lists possible parallels in engineered autonomous systems. However, whereas control in the biological system is carried out by the activity of interconnected

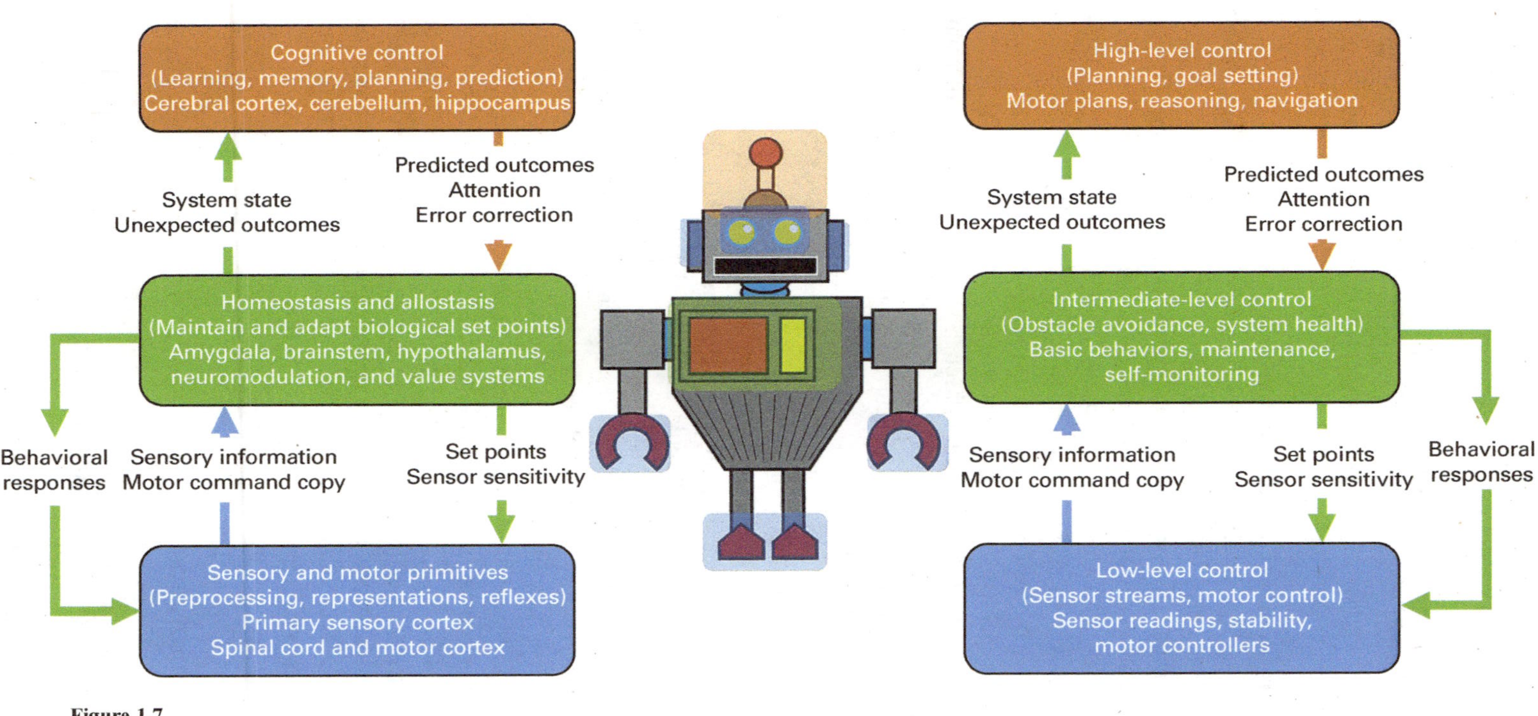

Figure 1.7
Schematic for self-monitoring systems in biology and in engineering. On the left are terms and regions derived from neuroscience. On the right are terms adapted from autonomous robots but could be applied to many embedded systems. Blue: low-level sensory processing and motor control. Green: intermediate-level homeostasis, maintenance, and monitoring. Orange: high-level planning, adapting, and goal-driven behavior. From Chiba and Krichmar (2020).

neurons, control in engineering systems can be carried out by algorithms. As discussed in the next chapter, many brain areas have circuits or topologies specific to their function. In both biology and engineering, the control system monitors sensors and drives mechanical systems by moving actuators.

1.3.2.4 Software architectures and approaches

There are many control architectures and approaches for robots. Typically, these are software programs to handle the intermediate and high levels of control. In the classic artificial intelligence architecture, sensory information is used to reason over a model of the world. An inference engine weighs the different possibilities and outcomes before deciding the appropriate action to take. This *sense, think, and then act* architecture was prevalent in early AI and cognitive architectures (Kotseruba & Tsotsos, 2018).

Neural networks are considered a part of AI today, although historically that was not always the case (Pollack, 1989). Neural networks take sensory information in their input layer of neurons, process that information through several inner layers of neurons, and then read an output layer of neurons to take an action. In some cases these artificial neural networks have feedback connections or recurrent loops that allow them to have history and interesting dynamics (Tani, 2016).

In the behavior-based approach, robots respond immediately and reflexively to environmental cues by using schemas (Arkin, 1998) or arbitration rules, as in the subsumption architecture (Brooks, 1986). The behavior-based approach was a reaction to the methodical sense, think, and act approach taken by the AI community. Behavior-based roboticists thought that effective robot control could be achieved faster and more efficiently by responding to the dynamics of the world without the burden of creating and maintaining a world model. In the schema approach, different schemas may represent different actions or sensory readings. They serve as templates that can be combined, similarly to adding vector fields, to create full suites of behavior. The subsumption architecture has a set of functions that can be triggered by sensory events. An arbitration scheme allows one function to interrupt or override others so that it can take control of the robot. These can be set up as a priority scheme or a hierarchy. For example, a robot's default behavior may be to move in a particular direction (i.e., wander module). If its sonar sensor detects something close, the robot may pause, wander, and move in the opposite direction of the nearby sensor reading (i.e., runaway module). If a camera detects an open corridor, this sets the robot's heading to move in the corridor's direction (i.e., explore). Once that heading is set, the wander module becomes active.

1.4 Neurorobotic Approach

Although neurorobotics is based somewhat on conventional robotics and AI approaches, it also has a strong foundation in behavioral neuroscience. Therefore, adopting ideas from the behavior-based approach is appealing to neuroroboticists. Although the brain does build models of the world to predict outcomes, the nervous system and organism are very responsive to environmental events. Reflexes at the spinal cord and attention to change operate rapidly, involuntarily, and often without conscious awareness. It is only after the event and our response has occurred that we become aware of what has just happened and take appropriate planning or corrective actions. An important observation of the early behavior-based robots was that their behavior looked much more natural than that of existing robots. Because neurorobotics is fundamentally interested in mimicking biological behavior, adopting such approaches at a low level or intermediate level of control makes sense.

The intermediate-level or high-level control of a neurorobot should follow some aspect of the nervous system. Therefore, neurorobotics is inspired by current methods in artificial neural networks. A major difference for the neurorobot controller is that its neural network should follow the architecture and dynamics of a real nervous system. It should have homologs to various brain regions. It typically has **neurons** and **synapses** (i.e., connections) that have some degree of biorealism. The neurorobot artificial nervous system is described with pathways connecting brain regions that have a variety of connectivity patterns, rather than with a feedforward architecture corresponding to the inner, hidden, and outer layers of neurons, as is prevalent in deep learning neural networks. These connections are often plastic and must support online learning. Unlike traditional AI systems, biological organisms do not have separate training and testing periods. They may have some innate capabilities but must learn through their own experiences. We like to say that no two neurorobots are alike. Even though they share the same robot body and have similar artificial brains, their experience shapes their behavior, which in turn can shape their nervous systems. Experience can change the way neurons respond over time and can cause rewiring of the synapses in the neural network.

Because of these details, it is important to provide some neuroscience background, which is the subject of the next chapter. Although brains are very complicated, understanding neuroscience is not too difficult once one gets through the arcane nomenclature. In many ways the brain's architecture and circuits have similarities to those in engineering and control systems. The complexity arises in the large number of elements and regions that make up even the simplest organism's brain. After reaching a basic understanding of neuroscience at a systems level, we will explore neurorobotic principles that follow the brain, body, and behavior model.

Notes

1. For a truly delightful video of these tortoises in action, see https://www.youtube.com/watch?v=lLULRlmXkKo.
2. https://observer.com/2015/08/stephen-hawking-elon-musk-and-bill-gates-warn-about-artificial-intelligence/.
3. https://en.wikipedia.org/wiki/Slaughterbots.

2 Neuroscience: Background for Creating Neurorobots

2.1 Introduction: The Need for Neuroscience

Neurorobotics attempts to closely follow the architecture and dynamics of the brain for two basic reasons: (1) testing a brain model under real-world conditions closes the loop by connecting brain, body, and environment. This is a more natural setting than a simulation in a highly constrained state space. Because the behavior can be observed under these more realistic environments while recording from all aspects of the artificial brain, neurorobotics may be informative of how brain activity can lead to complex behavior; and (2) biological organisms that have brains demonstrate what we call intelligent behavior and cognition. However, such capabilities have not been realized in artificial systems. Therefore, robotics and artificial intelligence may benefit from closely following neuroscience in designing a model. Moreover, the brain does not work in a vacuum. It is tightly coupled to the body in both its sensing and its actuation. Furthermore, the organism's brain and body are highly tuned to environmental cues and can alter the environment to suit the organism's needs. Because it senses and acts, the robot can be a surrogate body and the artificial neural network can provide the brain control of this body. But before we can make our own neurorobots, we need to review some concepts from neuroscience.

2.2 Neurons and Synapses

Brains are composed of neurons and synapses, as well as supporting cells known as glia and a vascular system to provide oxygen and nutrients. The number of neurons and synapses varies among organisms (see figure 2.1). However, all organisms with brains have neurons and synapses. Furthermore, all vertebrates (e.g., mice, cats, dogs, apes, humans, and elephants) have roughly the same types of neurons and synapses. Remarkably, despite the range in the size of these organisms, the sizes of neurons and synapses are similar across vertebrate species. However, the number of brain regions, neurons, and synapses can vary considerably from one organism to another.

Species	Neurons	Synapses
Nematode	302	10^3
Fruit fly	100,000	10^7
Honeybee	960,000	10^9
Mouse	75,000,000	10^{11}
Cat	1,000,000,000	10^{13}
Human	85,000,000,000	10^{15}

Figure 2.1
The number of neurons and synapses in different organisms. Adapted from http://en.wikipedia.org/wiki/List_of_animals_by_number_of_neurons.

2.2.1 Properties of Neurons

Neurons are highly specialized for generating electrical signals in response to chemical and other inputs (see figure 2.2). Neurons have a wide variety of shapes and sizes. Neurons contain a cell body, called the **soma**, which integrates inputs from the rest of the neuron. Typically, a neuron receives inputs through connections on its **dendrites**. Dendrites can have an elaborate branching structure that looks like a tree. The dendritic tree allows a neuron to receive inputs from many other neurons through synaptic connections. The output of a neuron is called an **axon**, which emanates from the soma and can traverse large distances to other parts of the brain or body. These axons connect with other neurons, often at the dendrite, through a **synapse**. Compared to computers, axon conductance of a signal is slow. Conduction velocities range from less than 1 m/sec to 200 m/sec.

To give an idea of scale, the motoneuron shown in figure 2.2*B* is several thousand microns wide, whereas the cerebellar Purkinje cell shown in figure 2.2*A* is about 100 microns wide. The Purkinje cell has a tree structure resembling a candelabra that can receive hundreds of thousands of synaptic inputs. A Purkinje cell turned on its side appears relatively thin. This allows inputs organized like cables to connect with packed rows of Purkinje cells.

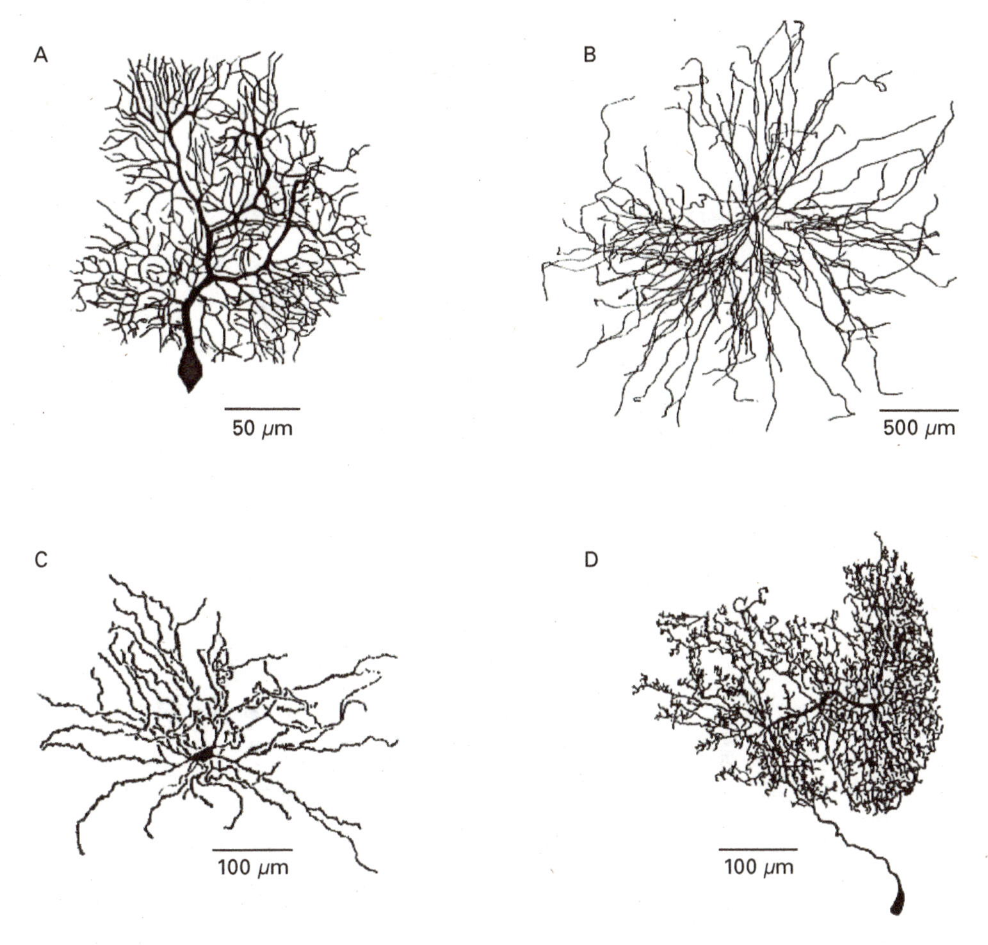

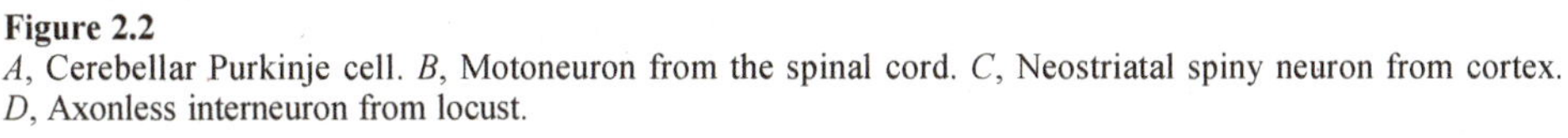

Figure 2.2
A, Cerebellar Purkinje cell. *B*, Motoneuron from the spinal cord. *C*, Neostriatal spiny neuron from cortex. *D*, Axonless interneuron from locust.

All this variety is due to fitting physical nodes and wires in a constrained three-dimensional space. More importantly, this variety is critically important for functionality. The **cerebellum** structure is designed for fine tuning of temporally sensitive information. The cortex has a columnar design that can be thought of as modules. This modular structure can be reused throughout the brain to handle different sensory inputs and drive different actions.

Neurons convey messages via an **action potential**, which is a rapid increase in **membrane potential** measured in millivolts, followed by a rapid decrease in membrane potential to a resting level (see figure 2.3). When neuroscientists and modelers talk about action potentials, they may say, "The neuron fired an action potential," or simply "The neuron

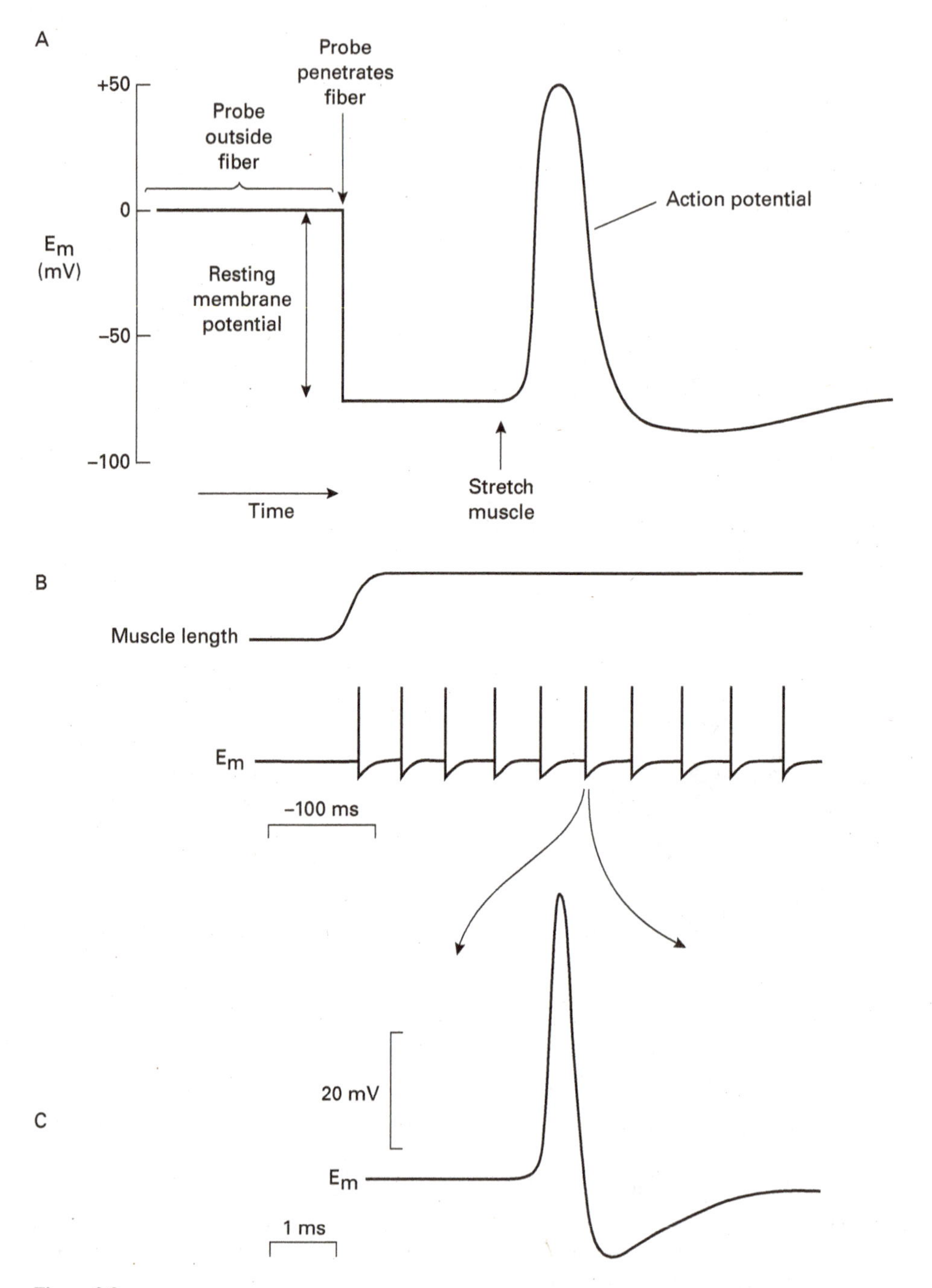

Figure 2.3
Generation of an action potential. *A*, The membrane potential, E_m, is measured in millivolts (mV). Because of differential chemical and concentration gradients of ions inside and outside the cell, the neuron is at rest around −70 mV. An input that causes the neuron to fire an action potential results in the E_m rapidly increasing to about +30 mV and decreasing to below its resting state. *B*, If this were a neuron in the spinal cord that gets muscle length as input, a stretching of the muscle would result in a train of spikes. *C*, One of the action potentials from the spike train.

fired" or "The neuron spiked." *Action potential*, *fired*, and *spiked* all refer to the same thing. An action potential is an all-or-none event. Once the membrane potential has increased above a threshold, due to ionic currents about which we do not go into detail here, the voltage increases and then decreases. The whole event takes about 1 ms to occur. Then there is a brief period during which the neuron cannot fire another action potential, known as the refractory period. Once the action potential has been generated, the signal travels along an axon with little attenuation until the axon reaches another neuron at the synapse. Therefore, we can think of an action potential as a binary event (0: the neuron did not fire or 1: the neuron fired). Compared to computers, which operate at gigahertz speeds, neural signaling is slow and sparse. A neuron that fires at 100 Hz is considered to be highly active. A typical cortical pyramidal cell fires at around 5–10 Hz.

Just as the signals of action potentials can be summarized as binary events, it is also common to summarize neuron activity in the form of overall firing rates rather than individual firing times. This is known as rate coding, in which the average firing rate of a neuron is usually measured in spikes per second or hertz (Hz). Appendix 2.A.1 provides a mathematical description of a neuron's firing rate.

2.2.2 Neuron Models

The action potential can readily be modeled with differential equations. Because these models capture the all-or-none aspect of an action potential, they are called spiking neurons. One of the simpler spiking neuron models is called the leaky integrate-and-fire (LIF) neuron. A LIF neuron integrates inputs, sums these inputs, and generates a spike if the sum is above some threshold. In the absence of input, the summation decays over time; that is, it leaks. Figure 2.4 shows the response of a LIF neuron. Appendix 2.A.2 provides a mathematical description of the LIF neuron and the parameters used to generate figure 2.4.

For most simulations, LIF neurons are adequate. However, for more realistic neuron models that better capture the variety of spike waveforms observed in different neurons, there are dynamical systems models, such as the Izhikevich neuron (Izhikevich, 2004) or LIF neurons that have extra terms (Brette & Gerstner, 2005).

Spiking neurons can capture the precise timing of spikes. However, because of their short time step, typically 1 ms, and the number of terms, they may be too computationally intensive to run on a robot in real time unless they are run on dedicated hardware designed for spiking neurons, known as **neuromorphic chips**. Neuromorphic chips, such as IBM's TrueNorth and Intel's Loihi, can take advantage of needing to calculate the neuron equation only when a spike is received, and needing to send a binary signal only when the neuron reaches threshold. In this chapter's case study we discuss a neuromorphic implementation of visual navigation. However, on conventional computers, spiking neurons carry a lot of overhead.

Mean firing rate neurons are a good alternative to spiking neurons. The time step for a rate neuron can range from several milliseconds to a quarter-second. In neuroscience terms, a mean firing rate neuron can be thought of as the average firing rate of a pool of

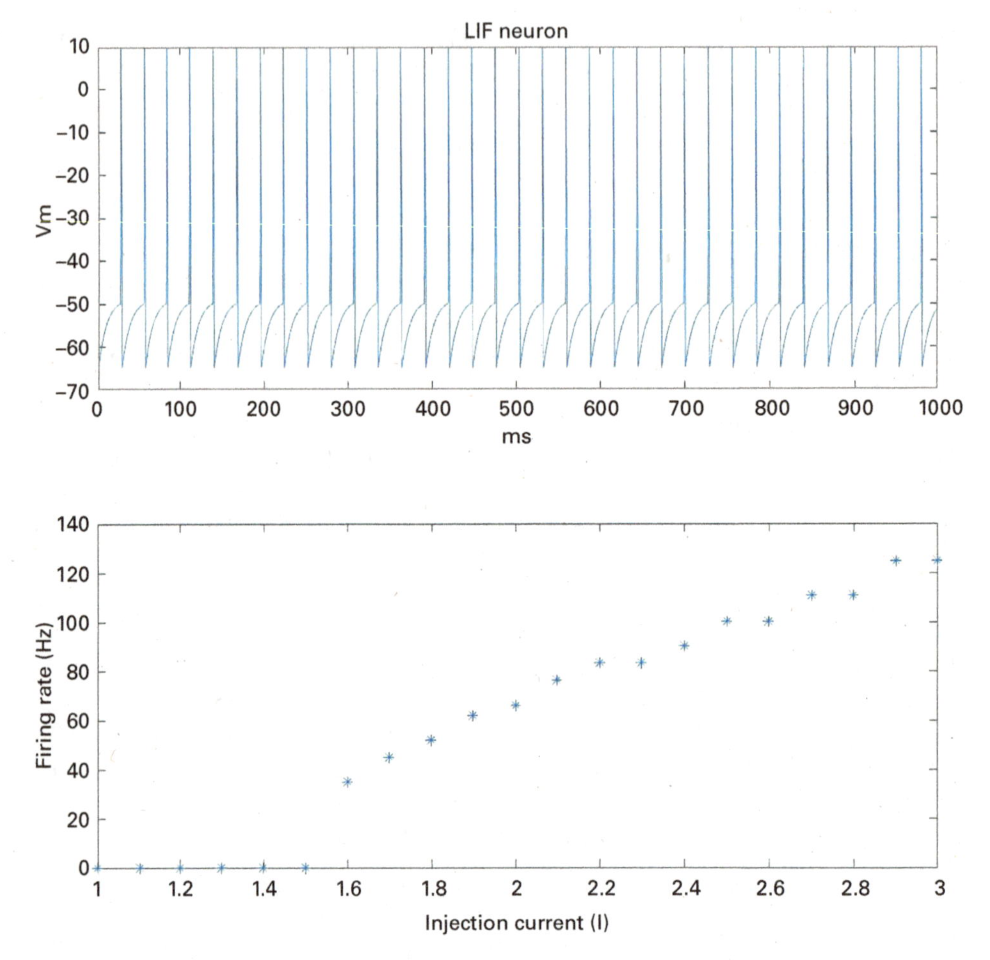

Figure 2.4
Leaky integrate-and-fire (LIF) neuron. *Top*, Response of the LIF neuron to an injection current, I, equal to 1.6. *Bottom*, Firing rate response over a range of injection currents.

neurons over tens of milliseconds. For example, a rate neuron using a time step of 16.7 ms could be used on a robot's visual system that has a camera with a 60 Hz frame rate without loss of data. Mean firing rate neurons take many forms. In artificial neural networks the sigmoid function, which looks like a forward-slanting *S*, and the threshold-linear function, also known as a rectified linear unit (ReLU), are commonly used. For computational neuroscience the sigmoid function is a good approximation of a firing rate response, and a radial-basis function (e.g., cosine curve or Gaussian function) is a good approximation of a sensory neuron response to input or a motor response for output. Appendix 2.A.3 shows examples of different rate functions and the equations to realize these curves.

2.2.3 Properties of Synapses

Synapses are connections between neurons. However, it is more complicated than that, and it is worth going into the history of their discovery. A brain slice viewed through a microscope looks like a tangled mess. For the untrained eye it is very difficult to separate the individual neuron's soma, dendrites, and axons from the supporting glial cells and structures. In the late nineteenth century Camillo Golgi invented a stain that darkened about 10 percent of the neurons, leaving the rest of the neuropil light colored, which made it easier to observe individual neurons. Golgi thought that the brain was a nerve net with cells physically connected to one another. Later, the great neuroanatomist Santiago Ramón y Cajal, while using Golgi's stain, came to a different conclusion. Ramón y Cajal proposed that there were gaps between neurons, which we now call synapses, that allowed distinct neurons to communicate with each other. In 1906, Golgi and Ramón y Cajal shared the Nobel Prize for their work in uncovering the structure of the brain.

Much later, electron microscopes enabled us to see the complex machinery of a synapse (see figure 2.5). Once an axon nears its target, it splits off into small ends known as axon terminals. The axon terminal, which sends a signal, is the **presynaptic** side of a synapse. On the other side, usually the dendrite, is the receiving side of the signal, called the **postsynaptic** side of the synapse. Between these is a space called the synaptic cleft. Some synapses, called gap junctions, are electrical, where current from the presynaptic neuron is received directly by the postsynaptic neuron. Gap junctions are relatively fast and are observed in neural circuits when precise timing is needed. However, most synapses are chemical; that is, the presynaptic side spews packets of chemicals, called **neurotransmitters**, across the cleft to be received by chemical-specific receptors on the postsynaptic side. In general, receipt of a neurotransmitter will cause the postsynaptic membrane potential to either increase or decrease. Although there is a wide variety of neurotransmitters with varying properties, at the most basic level they either increase or decrease the membrane potential. Thus, we usually describe synapses as excitatory or inhibitory.

If the postsynaptic neuron is excited enough that the membrane potential goes above the neuron's threshold, this postsynaptic neuron will fire an action potential. In essence, the postsynaptic neuron is now the presynaptic neuron. It may take several excitatory synapses to cause the postsynaptic neuron to fire. These signals are integrated at the soma. The signal input necessary for a spike can be a factor of synaptic event timing, distance of the synapse from the soma, the synaptic strength, and more. Appendix 2.A.4 describes how to model synapses for spiking neurons and mean-firing-rate neurons.

2.2.4 Synaptic Plasticity

The ability to learn and remember things is due mainly to changes at the synapse. The most basic idea is that if the presynaptic and postsynaptic neuron are active (i.e., fire action potentials) around the same time, the strength or efficacy of the synapses increases. This leads to the adage, "Neurons that fire together, wire together." This mechanism was predicted

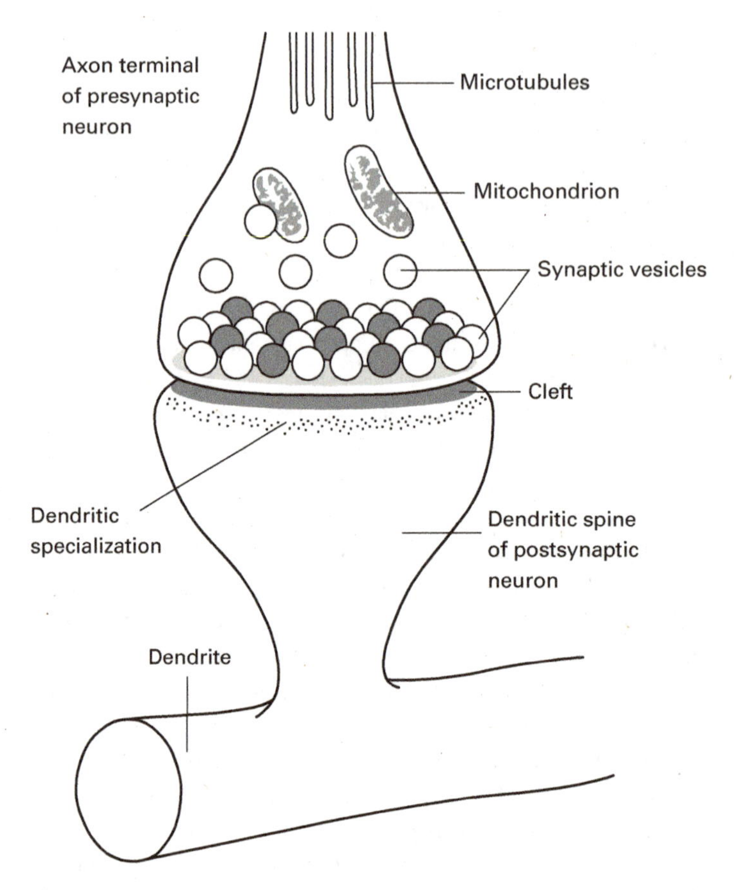

Figure 2.5
The synapse. On the presynaptic neuron, the axon terminates. Synaptic vesicles move to the edge of the axon terminal and release their neurotransmitters into the cleft. Receptors on the postsynaptic neuron receive the neurotransmitter, causing the membrane potential to either increase (excitatory synapse) or decrease (inhibitory synapse) the membrane potential of the postsynaptic neuron.

by neuroscientist Donald Hebb in the 1940s and became known as the Hebb rule. Initially, this mechanism was found in the **hippocampus**, a brain area known to be important for learning and memory. But, since then such synaptic plasticity has been found in many brain regions. The basic idea of the Hebb rule can be realized by delivering a strong rapid pulse to neurons in a brain slice, causing the strength of the synapse to increase and stay increased. This increase in synaptic efficacy is called **long-term potentiation** (see LTP in figure 2.6). However, equally important is being able to decrease synaptic strength when something needs to be unlearned. This can be achieved with a low-frequency pulse leading to **long-term depression** (see LTD in figure 2.6).

More recently, it was shown that Hebbian-like learning can occur due to the timing of presynaptic and postsynaptic spikes; this is known as **spike timing–dependent plasticity**,

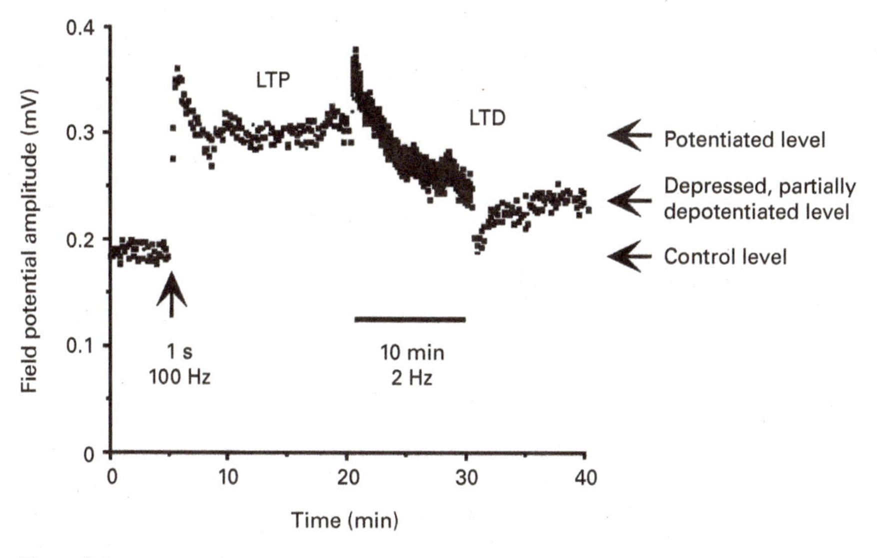

Figure 2.6
Long-term potentiation (LTP) can be achieved with a strong rapid pulse of activity. Long-term depression (LTD) can be achieved with a slower pulse of activity.

or STDP (Bi & Poo, 1998; Markram et al., 1997). If the presynaptic spike (neuron A) precedes the postsynaptic spike (neuron B), there is LTP (see pre-before-post in figure 2.7). This makes sense in that if neuron A caused neuron B to spike, these two neurons are causally related and the connection between them is strengthened. If the timing is reversed (neuron A fires after neuron B), there is LTD, which also makes sense since these two spiking events are not causally related, and the connection is weakened (see post-before-pre in figure 2.7). The amount of LTP or LTD is related to how closely in time these spike events occur. Typically, closer timing of these spike events results in a stronger effect.

Synaptic plasticity is the main ingredient for achieving learning and memory in brains and in models. There are many variations of these learning rules, some of which we explore in greater detail throughout the book.

2.3 Systems Neuroscience

The brain is called a nervous system because it is a system of interconnecting regions with different functions and specialties. There are separate sensory streams for hearing, touch, vision, and motor outputs specific to moving different parts of the body (e.g., eyes, head, arms, and legs). There are also regions important for different types of memories, such as **declarative memory** (i.e., places and events) or **procedural memory** (i.e., skills). There are brain areas that are important for short-term or working memory, which operates over seconds and minutes, and brain areas that are important for long-term memory that can last

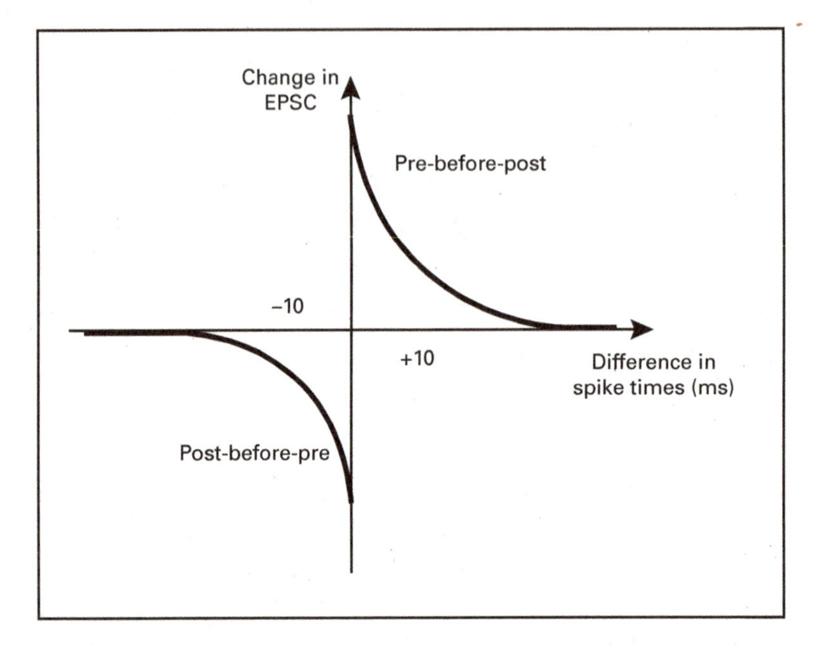

Figure 2.7
Spike timing–dependent plasticity (STDP). Long-term potentiation (LTP) and long-term depression (LTD) are measured by changes in the excitatory postsynaptic current (EPSC). LTP occurs if the presynaptic spike precedes the postsynaptic spike. LTD occurs if the postsynaptic spike precedes the presynaptic spike. The amplitude of change is larger if the events are closer in time, and the amplitude of change falls off exponentially with time.

from several minutes to a lifetime. The brain has hubs where all this information is received and integrated (Sporns, 2010). Despite all these different functional areas, the brain is highly connected at both the micro and macro scale. The brain is a **small-world network** in which it is only a few short hops, via synaptic connections, from one region to another (Bassett & Bullmore, 2017; Watts & Strogatz, 1998). Figure 2.8 shows the two extreme networks from completely ordered (regular) to completely disordered (random). A small-world network lies somewhere between these extremes and the brain follows that general pattern.

The field of systems neuroscience looks at the brain in a similar way to how an engineer looks at a control system, where there are inputs, controllers, state information, and outputs. Because a robot is basically a control system on wheels (or legs), it is important for neuroroboticists to create their designs with a systems neuroscience point of view.

Figure 2.9 shows the main anatomical structures in the nervous system. The **central nervous system** (CNS) consists of the brain and the spinal cord. The brain includes all the various neural regions within our skull, such as the neocortex, brainstem, and cerebellum. The CNS continues from the brain to the spinal cord, where different nuclei control movements and take in sensory information from the muscles, joints, and skin. The **peripheral nervous system** (PNS) handles sensory information from the internal (i.e., within the

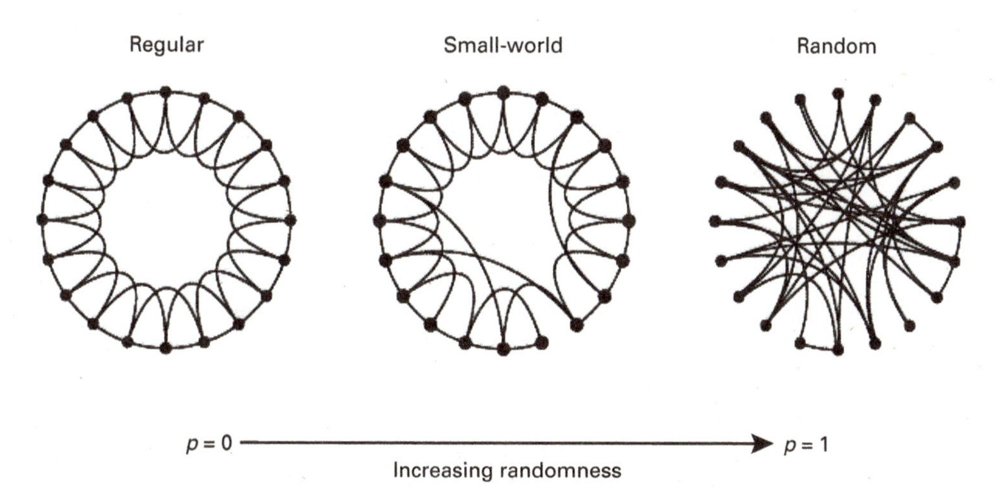

Figure 2.8
The Watts-Strogatz model and the generation of small-world networks. The model begins with a regular lattice network in which each node is placed along the circumference of a circle and connected to its k nearest neighbors. Edges are rewired uniformly at random such that at zero probability, $p = 0$, the network is still a lattice and at 100 percent probability, $p = 1$, the network is random. Intermediate values of p result in small-world networks with local clustering and short path lengths. Adapted from Watts and Strogatz (1998).

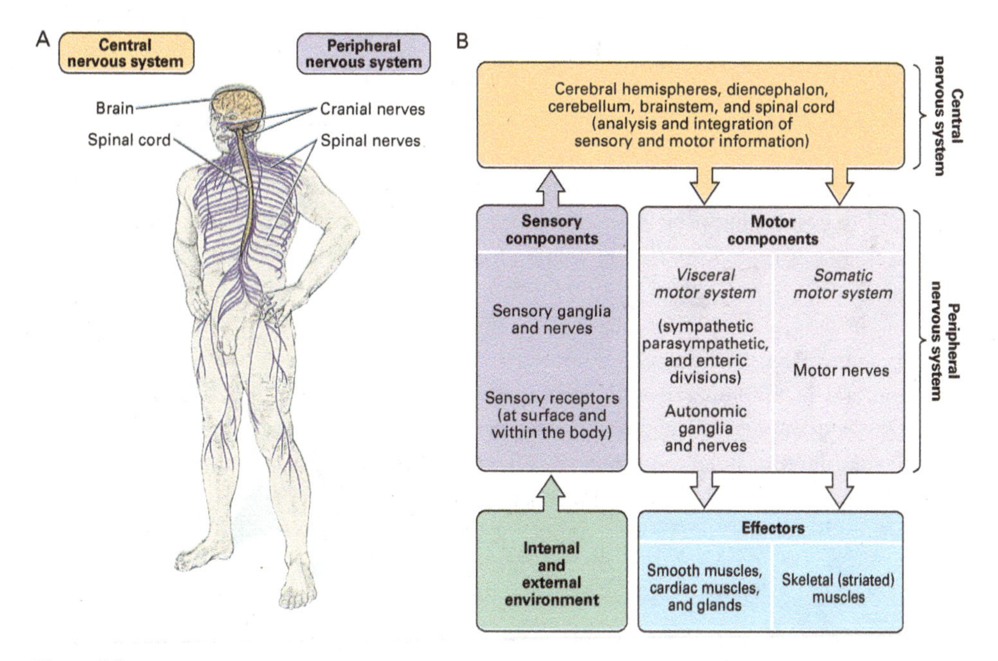

Figure 2.9
The major anatomical components of the nervous system and their functional relationships. *A*, The division between the central nervous system and the peripheral nervous system. *B*, The different divisions of the nervous system. From figure 1.12 in Purves et al. (2017).

body) and external (i.e., outside the body) environments. The PNS also includes the motor system with neurons to move parts of the body (somatic motor system), and neurons to control heart rates, digestion, and other bodily functions (visceral nervous system).

2.3.1 Cortical Structures

Within the brains of mammals is a folded structure known as the **neocortex**, which consists of repeating columnar structures, and is a six-layer cellular structure thought to evolve later in the cortex of mammals. Note that the hippocampus is sometimes called the *archicortex* since is phylogenetically older than the neocortex. Early cortical processing of sensory input takes place in the primary sensory cortex (see figure 2.10*A*). The primary sensory cortex extracts early features of stimuli specific to a given sense. For example, the primary visual cortex has neurons that respond specifically to visual features. Neurons in the primary auditory cortex respond to specific sounds. The primary motor cortex indirectly and directly innervates muscles to cause movement. Association cortices are multimodal; they integrate multiple sensory signals from the primary and secondary sensory cortex (e.g., vision and hearing) and generate activity in the motor cortex. A remarkable feature of the cortex is that despite its wide range of specificity and functionality, the cortex comprises a series of repeating columnar structures (see figure 2.11*B*). The main differences among these structures are the inputs to and outputs from them.

For example, the different columns in the motor cortex and the somatosensory cortex represent different parts of the body (see figure 2.11). In the case of the motor cortex (figure 2.11*A*), stimulating different brain areas will cause different parts of the body to move due to descending projections to the spinal cord and eventually muscles. In the case of the somatosensory cortex (figure 2.11*B*), touching different parts of the body will send signals via afferent projections from the periphery to the spinal cord and up to the cortex, resulting in activity in specific areas of the somatosensory cortex. Note the different sizes of body maps. Because figure 2.11 is based on the primate, there is overrepresentation of the hand and face compared with other body areas. This suggests that experience and necessity (both ontogenetic and phylogenetic) shape the structure of these representations. Primates are highly social, and thus there is a need for strong sensory and motor representations associated with the face. Primates also use their hands regularly and thus require fine sensory resolution and motor control for manipulating objects.

2.3.2 Sensory Systems

For each of our senses (hearing, smell, taste, touch, and vision), there are pathways from the sensory receptors to the cortex in the CNS. The pathways start at the periphery and after just a few synapses reach the **thalamus** and from there proceed to the cortex. An exception to this stream of information is the sense of smell, in which olfactory receptors bypass the thalamus on their way to the olfactory cortex.

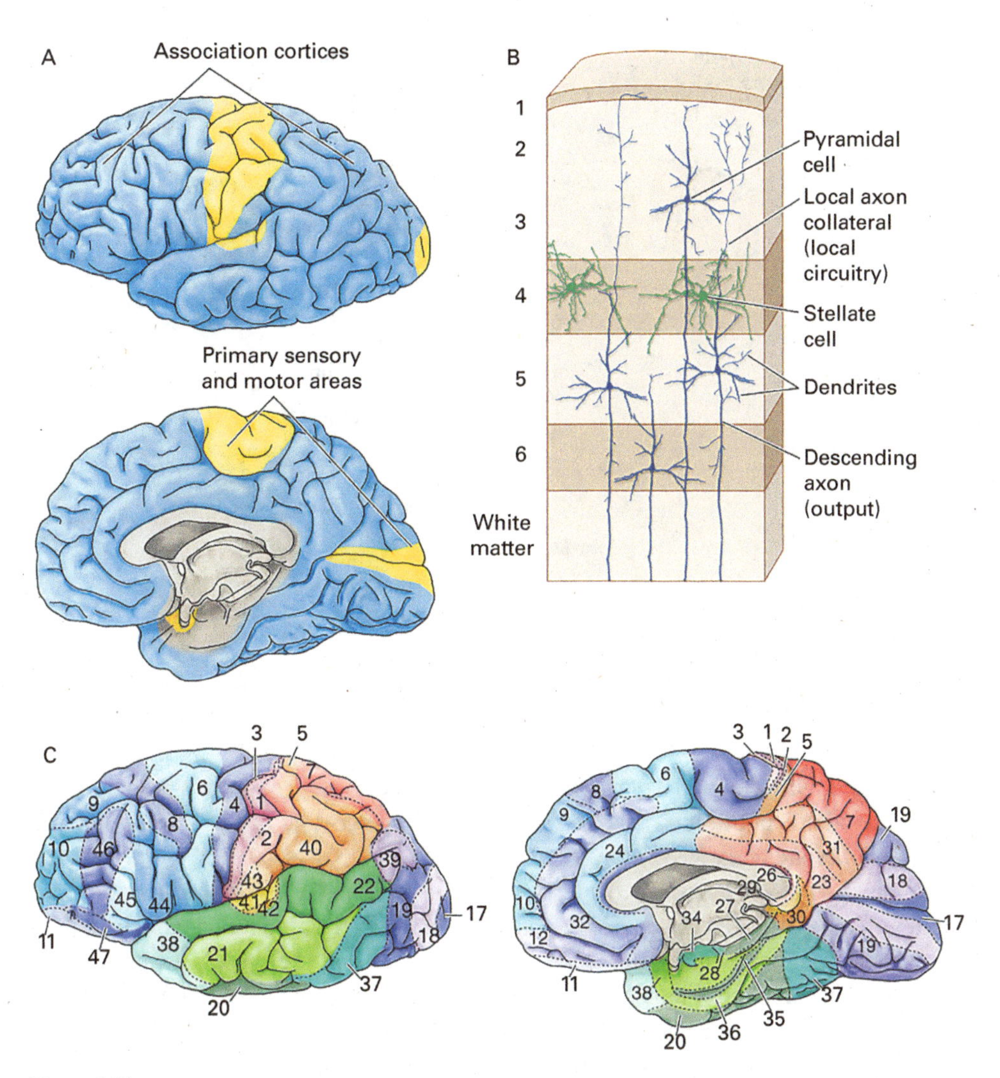

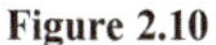
Figure 2.10
Structure of the neocortex. *A*, Association cortex shown in blue. Primary sensory and motor cortex shown in yellow. *B*, A cortical column. *C*, Lateral and sagittal views of the cortex with Brodmann area numbers. From figure 27.1 in Purves et al. (2017).

It would take several chapters to describe the variability and features of each sensory system, so we concentrate on vision for several reasons. One reason to focus the discussion on vision is that the expansion from simple to complex features in the visual system has similarities to other sensory streams. Compared with other sensory systems, the various features (lines, colors, textures, etc.) that make up a visual scene are relatively easy to

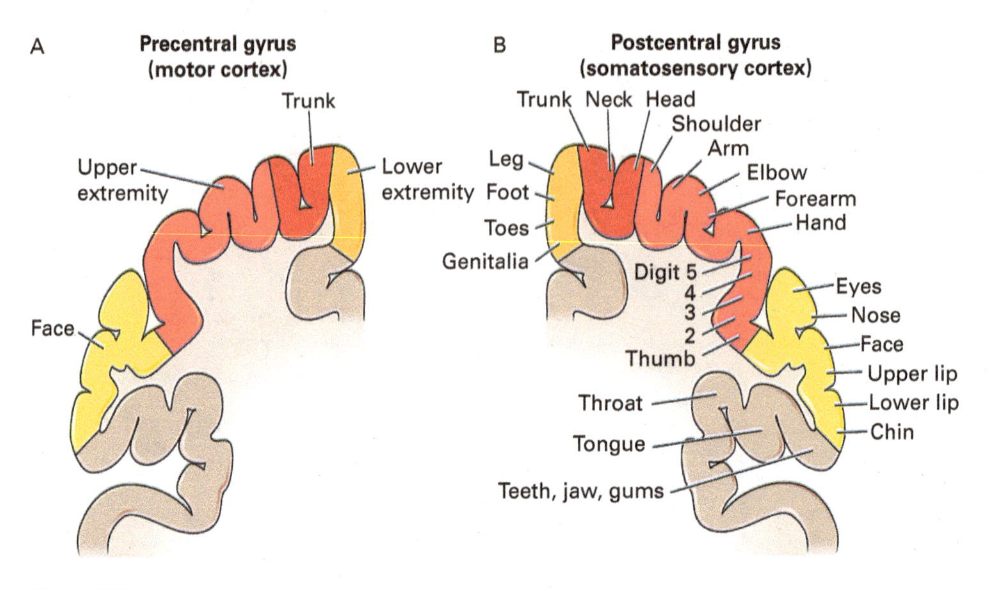

Figure 2.11
Motor and touch maps in cortex. Both the somatosensory and motor cortex have maps of the body. *A*, Motor cortex. Stimulating different parts of the motor cortex will cause movements of specific parts of the body. *B*, Somatosensory cortex. Touching different parts of the body will cause activity in different parts of the somatosensory cortex. From figure 1.18 in Purves et al. (2017).

comprehend. Furthermore, vision is commonly used for sensing in robotics. Cameras with high resolution are readily available. The models and processing of visual information in neurorobotics are often inspired by vision in the brain.

Figure 2.12 shows the pathway of the visual stream from the eye to the primary visual cortex. Light reaches the eye and activates photoreceptors in the retina, which respond to contrast, luminance, and color. The retina itself does some very sophisticated preprocessing of this light information before sending that information to the lateral geniculate nucleus (LGN) in the thalamus. Note in figure 2.12 that many of the signals from one side cross over to the other side (contralateral), whereas some other signals stay on the same side (ipsilateral). The LGN responds to spots of light surrounded by darkness (on center, off surround) or just the opposite (off center, on surround). These neurons can detect changes in contrast. From there, the LGN projects to the primary visual cortex, which is often called the striate cortex or V1. V1 responds to short, oriented segments that can provide information about the shapes of objects. Before moving further down the cortical stream, note that close examination of figure 2.12 shows a projection from the retina to the **superior colliculus**. The superior colliculus controls rapid eye movements known as saccades and can point the eye to objects of interest. Such movements are important for visual attention.

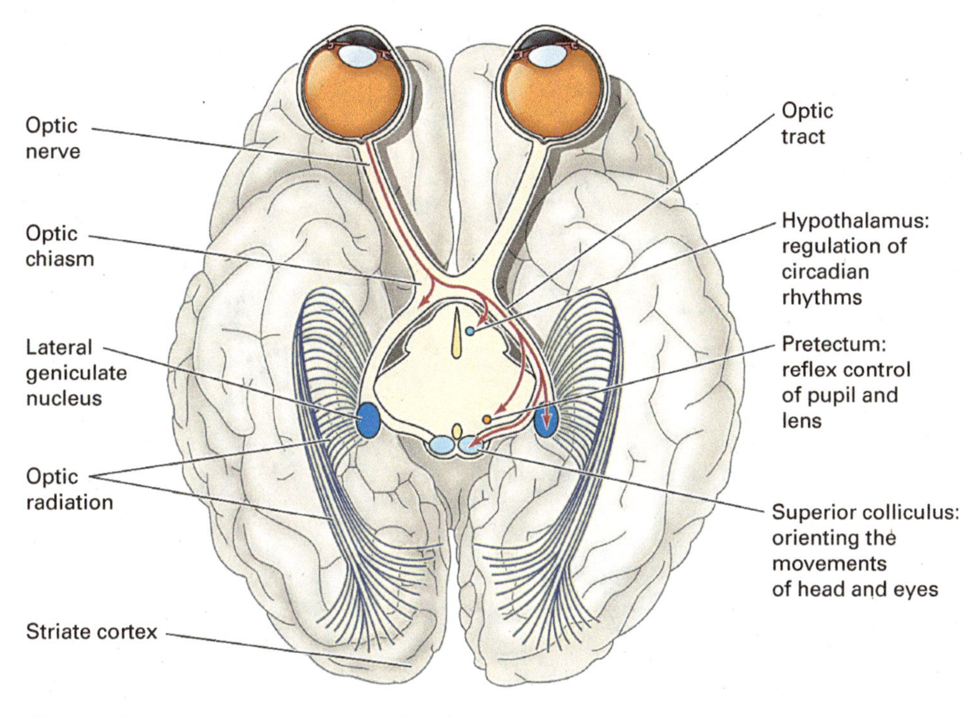

Figure 2.12
Visual stream. Major pathways from the eye to primary visual cortex. From figure 12.1 in Purves et al. (2017).

The visual stream continues from V1 to several cortical areas (see figure 2.13). As information proceeds along the stream, the encoded features become more complex by combining earlier features (Kravitz et al., 2013). In addition, the area sampled by each neuron, called its **receptive field**, increases in size. This expansion allows for neurons to take in more of the scene. For example, a V1 neuron may respond to a specific shape of a small area in a picture, and V2 and V4 may respond to the curve and color within a larger part of the picture. A neuron in the inferior temporal cortex (TE in figure 2.13) may combine these features to respond to an animal invariant of the size, position, and view of that animal in the visual scene. Figure 2.13*A* highlights another important aspect of visual processing in the brain. Roughly from V4 onward, the stream splits into two streams; the ventral stream, which encodes the object type (i.e., what), and the dorsal stream, which encodes the object location (i.e., where).

2.3.3 Motor Systems

The main goal of the brain is to promote action, and thus there are extensive brain areas dedicated to movement. Movements can be reflexive or involuntary in response to some

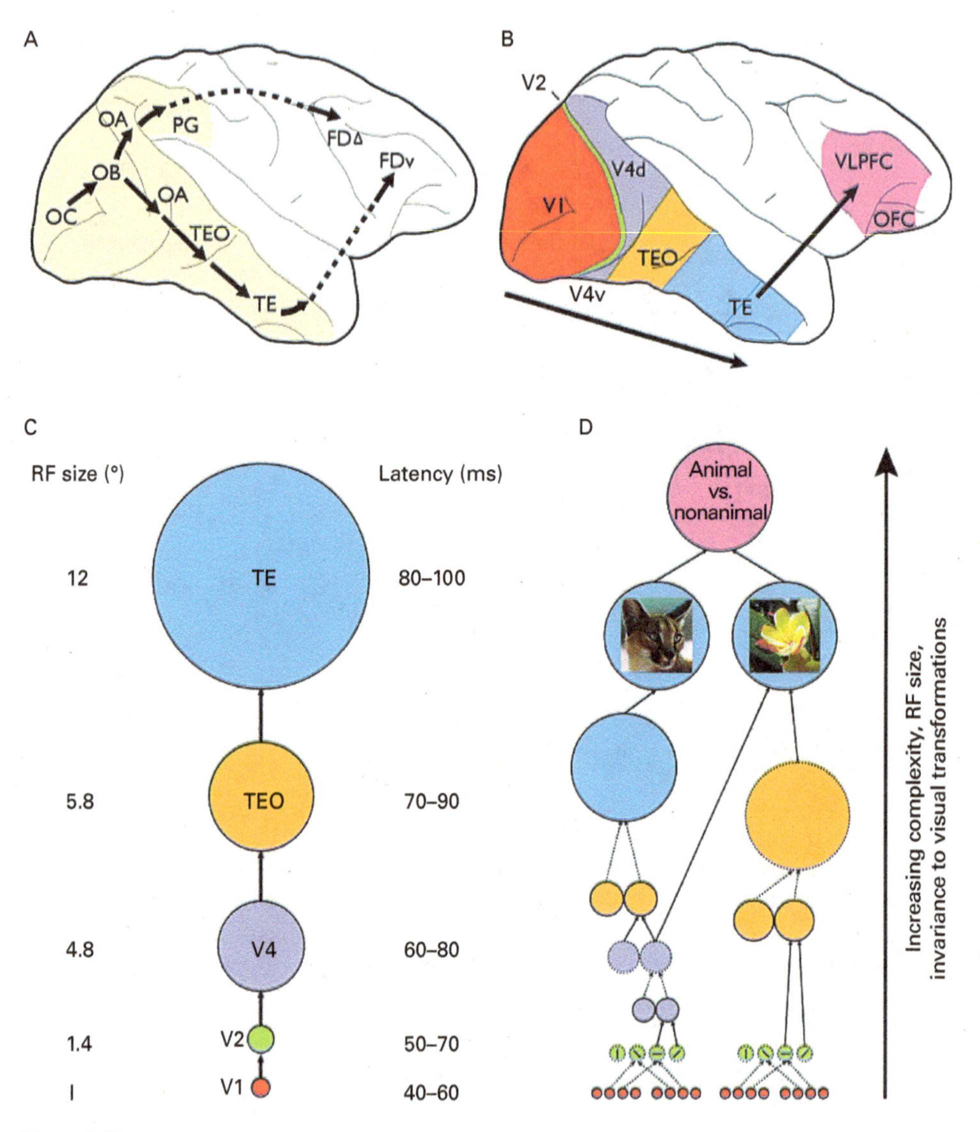

Figure 2.13
Visual cortex streams. *A*, Dorsal and ventral pathways in a macaque monkey. The ventral pathway is a stream projecting from striate cortex to area TE in the inferior temporal cortex. The dorsal pathway is a stream projecting from striate cortex to area PG in the inferior parietal cortex. The ventral pathway supports object vision (i.e., what), and the dorsal pathway supports spatial vision (i.e., where). *B*, Schematic of the ventral pathway that lies along the lateral surface of the macaque brain and its projection frontal cortex. *C*, Schematic of serial information flow from V1 to inferior temporal cortex (TE). The size of each circle reflects the average receptive field (RF) size of neurons in that region from recent recordings. *D*, Schematic of a version of the HMAX model of object recognition in the ventral stream. Adapted from Kravitz et al. (2013).

stimuli. For example, stepping on a sharp object with one's left foot causes the left foot to automatically lift up and the right foot to come down to prevent a fall. This reflex circuit occurs between the legs and the spinal cord and does not involve the brain. The brain would be too slow to respond to such an event without incurring damage. Another example of an involuntary movement involves bottom-up attention. Suppose that you see something large in your periphery. Your eyes will reflexively move to center on this potentially dangerous object. As mentioned in the previous section, this is a short circuit from the eye to the superior colliculus and then to the brainstem to control the movement of eye muscles.

Movements can be planned or voluntary. In a chess match, you may plan several movements in your mind before choosing what you think is the best movement. Once that choice has been made, there is a preparatory signal in the premotor cortex, which drives specific hand and arm areas in the motor cortex, which goes through the spinal cord and eventually to the muscles controlling the arm, wrist, and fingers. Prior to the movement many brain areas, including the frontal and parietal cortex, may have been involved in visualizing the movements and assessing different options.

Figure 2.14 shows some of the brain areas involved with movement. The motor cortex and frontal cortex are important for the planning, initiating and execution of voluntary movements. The brainstem is important for both voluntary and involuntary movements, which includes posture control and rhythmic movements such as locomotion or chewing. Signals from the cortex and brainstem descend to the spinal cord to innervate muscles causing movement. Proprioceptive input from the movement (e.g., joint position, torque, and load) is received at the spinal cord causing adjustments to these movements.

Figure 2.14 shows two different brain areas, the **basal ganglia** and the cerebellum, which further contribute to motor functionality. The basal ganglia can initiate or suppress movements through a highly organized structure seemingly designed for action selection. Once selected, these actions can be sequences of movements or stereotypical motor programs. The cerebellum has a highly organized structure for organizing movement, exercising fine motor control with precise timing, and adapting movements when there are errors.

2.4 The Neurorobotics Approach to Systems Neuroscience

Neuroscience and cognitive science are taught by separating the fundamental functions that the brain performs into different topic areas. In fact, this chapter has borrowed liberally from the sixth edition of *Neuroscience* edited by Dale Purves and colleagues (Purves et al., 2017). This book and most other textbooks on the subject distinguish sensory systems from motor systems through different chapters on memory, attention, executive control, and emotion. In reality, the organism does not have separate boxes

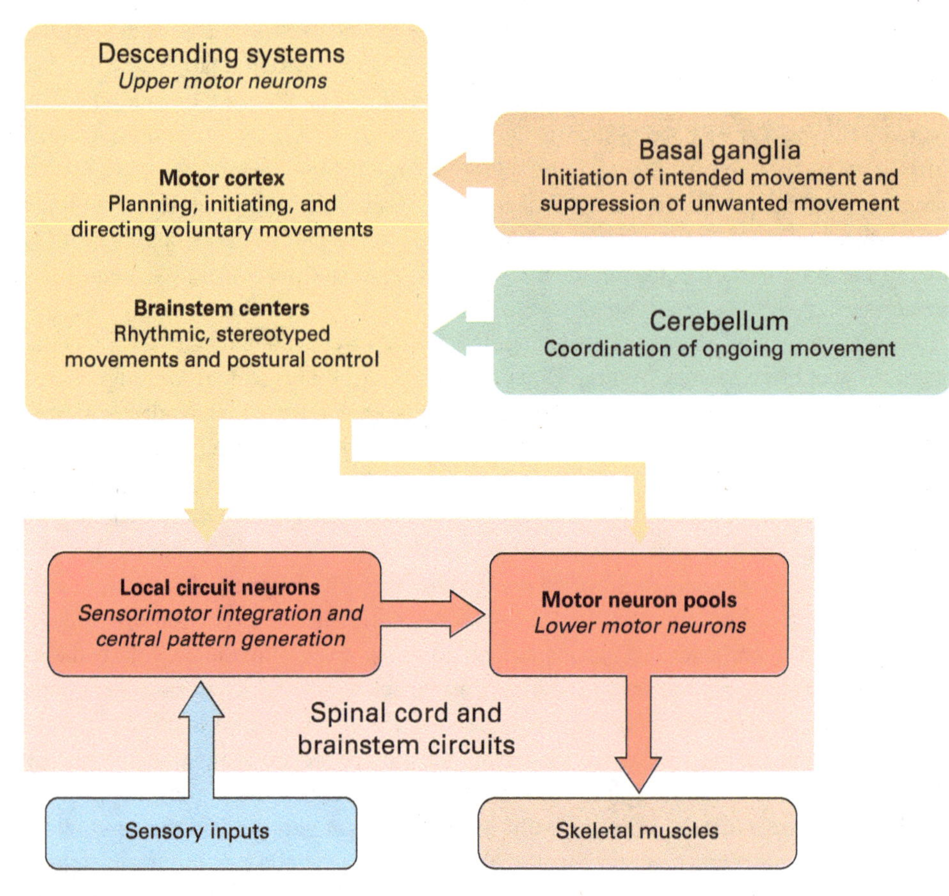

Figure 2.14
Brain areas involved in motor control. Adapted from figure 16.1 in Purves et al. (2017).

for each of these fundamental functions. Of course, there are brain areas that are more motor than sensory. However, as mentioned previously, the brain is a small-world network. Motor areas strongly and bidirectionally interact with sensory areas. Attention is important for memory, and our memories can dictate what we attend to. The same can be said for emotion. All these factors go into the day-to-day, moment-to-moment decisions made about what actions to take. Moreover, the brain's function is to serve the body's needs.

Because of its holistic nature, neurorobotics has the potential to address these issues. A neuroroboticist needs to take into consideration the brain (e.g., an artificial nervous system), the body (e.g., a robot that can carry out some behavior), and the environment (e.g., the portion of the world that the robot samples with its sensors and acts upon with its actuators).

In essence, the neuroroboticist is acting like a field biologist by observing the robot's behavior in the wild, with the added benefit of being able to keep a complete record of the robot's brain activity during this behavior. Even if the artificial brain is designed with separate motor, sensory, and decision areas, the interaction between these areas cannot be studied separately because the robot cannot isolate one from the other when behaving in the real world.

There are also advantages and lessons to be learned by having the model agent actively engaged in the world. Such embodiment can facilitate seemingly difficult functions. For example, figure-ground segmentation is a difficult computer vision problem. However, if the agent can manipulate objects in the world, the object can readily be separated from the background (Fitzpatrick & Metta, 2003). Figure 2.15*A* shows an experiment in which the robot randomly moves its arm across a table. As soon as it touches and moves an object, it becomes obvious to the robot that this object is separate from the table. Our visual systems are highly tuned to movement and our own actions can generate a wealth of salient sensory information.

Another example in which the embodiment facilitates processing is invariant object recognition. The Darwin VII experiment shown in figures 2.15*B* and 2.15*C* is an example of how viewing a scene as a stream of information, rather than separate static images, can lead to invariant categories for visual objects (Krichmar and Edelman, 2002). Darwin VII moved around in search of play blocks to pick up. Its camera sent a stream of video frames to its visual system. Darwin VII's primary visual cortex (VA_PH in figure 2.15*C*) responded quickly to contrast changes. VA_P projected to a simulated inferior temporal cortex (IT in figure 2.15*C*) with plastic connections that changed similarly to the Hebb rule discussed previously. Because the IT neurons had some persistence, stable representations of an object were activated as Darwin VII approached and picked up the play block. This natural way of viewing the world led to a perceptual object category that was invariant to size, position, and orientation. Furthermore, when the order of the video frames was randomized, invariant object recognition was impaired.

2.5 Case Study: Visual Navigation in Insects and Mammals

Organisms ranging from insects to humans must travel through dynamic, cluttered environments without crashing into things. We do this with such ease that oftentimes we are not aware of it, as when we move through a crowd or hike through a forest. In general, vision is used to routinely scan the environment, avoid obstacles, and approach goals. These visual signals can rapidly convey important information to our motor systems, resulting in fluid, seamless movement through the world.

In the case of the honeybee, a simple centering response that balances the **optic flow** on both sides of the bee allows it to fly through tight spaces (Srinivasan et al., 1991;

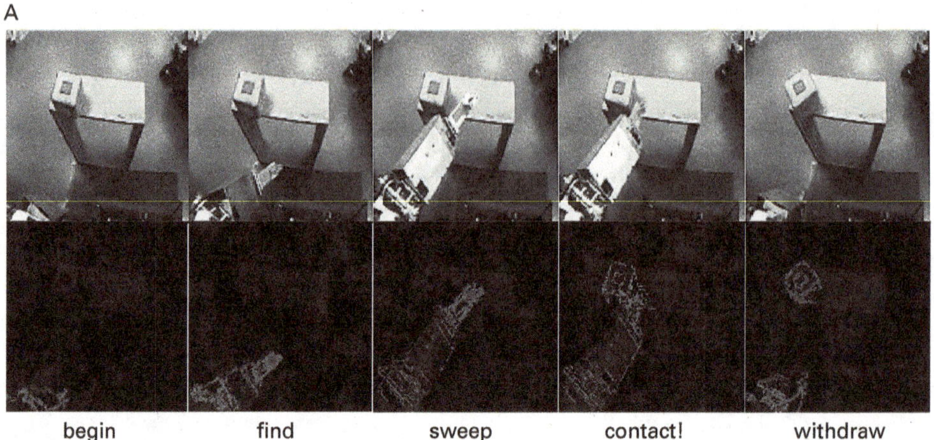

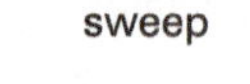

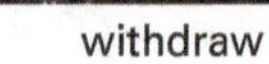

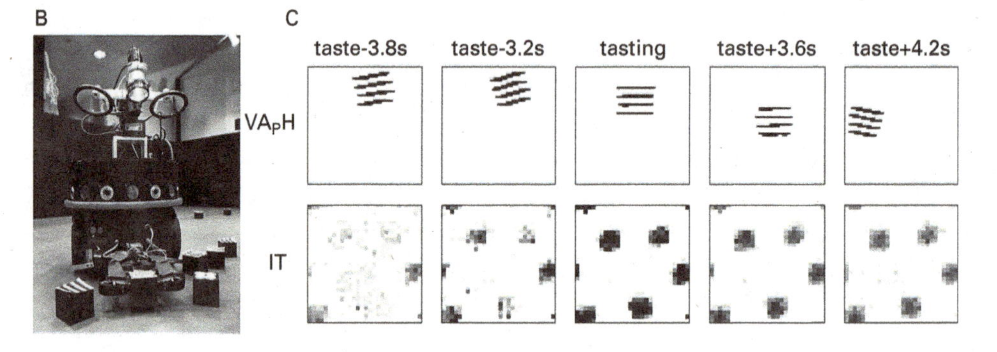

Figure 2.15
A, The upper sequence shows an arm extending into a workspace, tapping an object and retracting. This is an exploratory mechanism for finding the boundaries of objects. The lower sequence shows the shape identified from the tap using simple image differencing. From (Fitzpatrick & Metta, 2003). *B*, Darwin VII Brain-Based Device. Darwin VII's gripper picked up and examined blocks with different visual patterns. *C*, Invariance with respect to position, scale and rotation emerges from a persistent pattern of activity in area IT as the pattern of activity in the VA$_P$H areas moves sequentially across Darwin VII's field of vision. *B* and *C* adapted from Krichmar and Edelman (2002).

Srinivasan & Zhang, 1997). Optic flow refers to the apparent motion across visual receptors due to moving objects or to one's own movements. Srinivasan and colleagues showed that bees balance the left and right flow fields by having bees fly through a tunnel with different grating patterns. If the flow between left and right sides were balanced, the bees flew down the center (see figure 2.16*A*). However, when one of the gratings was in motion, bees flying in the same direction as the moving grating tended to fly closer to the moving grating because the optic flow was greater on the stationary grating (see figure 2.16*B*). However, if the grating was moving in the opposite direction as the bee, which meant the

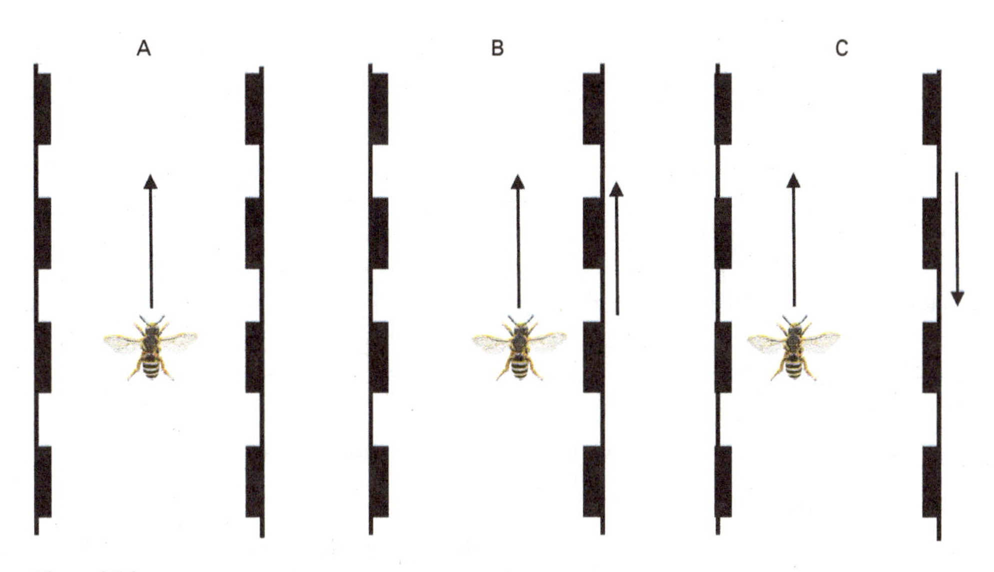

Figure 2.16
Balanced optic flow as a visual navigation strategy. A bee is shown flying down a corridor with black and white stripes. *A*, When the stripes are stationary, the bee flies down the center of the corridor. *B*, When the stripes on the right side are moving in the same direction of the bee, the bee moves towards the right to balance the flow. *C*, When the stripes on the right side are moving in the opposite direction of the bee, the bee moves toward the left to balance the flow.

moving grating side had greater optic flow, the bee would move toward the stationary side to balance the flow fields. This showed that flying bees estimate the distances of surfaces in terms of the apparent motion of their images. We can model this behavior with a straightforward balancing equation:

$$turnRate = G\frac{(OF_{\text{left}} - OF_{\text{right}})}{(OF_{\text{left}} + OF_{\text{right}})},$$

where OF_{left} and OF_{right} are the optic flow on the left and right fields, respectively, of vision, and G is a scalar gain term. If the *turnRate* is positive, the agent should move to the right, and if *turnRate* is negative, the agent should move to the left. There is evidence that humans also use this strategy (Kountouriotis et al., 2013).

In this case study, we will discuss two neurorobotic examples of visual navigation that use an optic flow balancing strategy. One mimics the insect neural circuit and the other simulates the mammalian visual cortex. Box 2.1 describes a simulation of visual navigation using the balancing strategy that can be tried out using the Webots simulator.

2.5.1 Trajectory Stabilization Using Optic Flow from Event-Based Sensors

In this first example of visual navigation, an optic flow strategy was implemented on a neuromorphic processor that used event-based vision sensors instead of cameras. Before describing the neurorobotic study, we briefly explain what we mean by *neuromorphic processor* and *event-based sensor*. Neuromorphic processors typically use spiking neurons and can decrease power by not computing between spike events. Sometimes the spiking neuron algorithm is implemented in software that runs on the processor. Other times the spiking neuron is implemented in the hardware itself. The neurons can be processed in

Box 2.1

A Webots simulation of how the balancing equation can be used to smoothly navigate down a corridor is provided in the BeeNavigation folder on GitHub: https://github.com/jkrichma/NeurorobotExamples/.

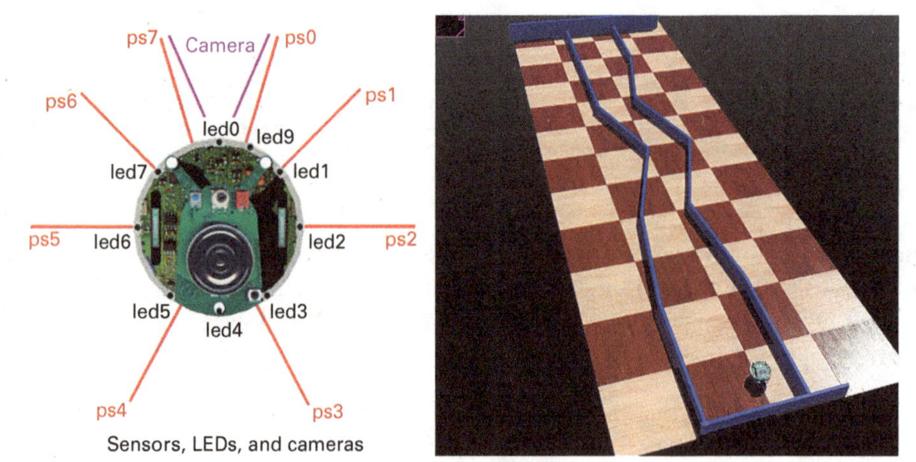

Figure 2.17
Left: E-Puck robot has eight proximity sensors (ps0–ps7) that measure distance to the walls. The range of the proximity sensors in the robot node was increased by a factor of ten to provide obstacle information further from the walls. *Right*: The e-Puck is at the bottom of the image and navigates up the maze depicted by the blue walls.

The robot should smoothly move down the center of the corridor (figure 2.17). How smoothly depends on its gain parameter. To measure the smoothness of its trajectory, we can calculate the coefficient of variation (CoV), which is the standard deviation divided by the mean $= \sigma | \mu$. CoV measures the dispersion of data points. For example, figure 2.18 shows the CoV of the turn rate from simulated bee navigation runs with different velocity gains. Note how the amount of turning increases as the gain increases. Less turning corresponds to a smoother trajectory.

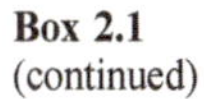

Box 2.1
(continued)

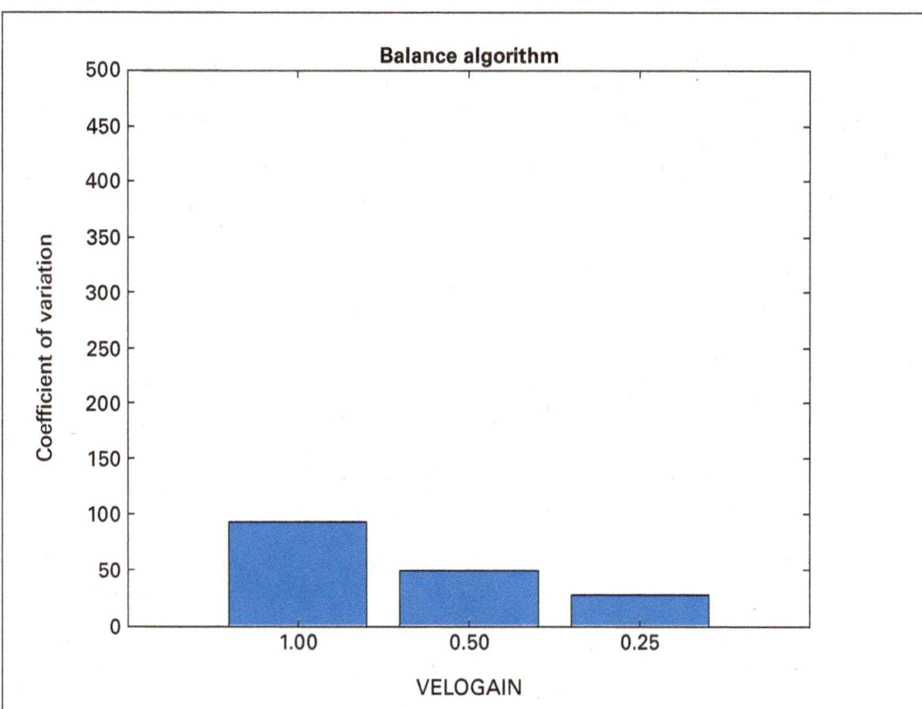

Figure 2.18
The effect of velocity gain on navigating using a balance equation. The chart shows the coefficient of variation of the turn rate.

Try different velocity gain values and record the robot's performance. Calculate the CoV for the different parameter settings. Change the maze configuration. Try replacing the balance equation with a simpler obstacle avoidance algorithm (e.g., only turn when very close to the wall). How do these changes impact behavior?

parallel and need to be sampled only when there is a spike event. When a spike event occurs, all that is needed to send to other neurons is the ID of the neuron that spiked and when that neuron spiked. Event-based vision sensors work in a similar fashion. If a pixel changes in contrast, then a spike event occurs for that pixel. If the scene is static, then there is no information to send. This is much different from a standard camera, which would send a full frame of pixels at 60 Hz. In this way an event-based vision sensor can respond much more rapidly than a conventional camera and operate at lower power. Because neuromorphic hardware has low size, weight, and power and uses event-driven, massively parallel, and distributed processing of information, it is ideal for autonomous robots in situations with limited power resources and limited Internet connectivity.

To investigate how trajectory stabilization could be realized with neuromorphic processing, Galluppi and colleagues (Galluppi et al., 2014) embedded a SpiNNaker neuromorphic chip and two embedded dynamic vision sensors (eDVS) on a robotic platform, shown in figure 2.19.

Similarly to the honeybee experiment, the robot had to traverse a corridor with vertical gratings (see figure 2.20*A* and 2.20*B*). Also like the honeybee, the robot attempted to stay in the center of the corridor by balancing the optic flow (OF), with raw visual information provided solely by two laterally mounted eDVS sensors (figure 2.19). The upper-left panel

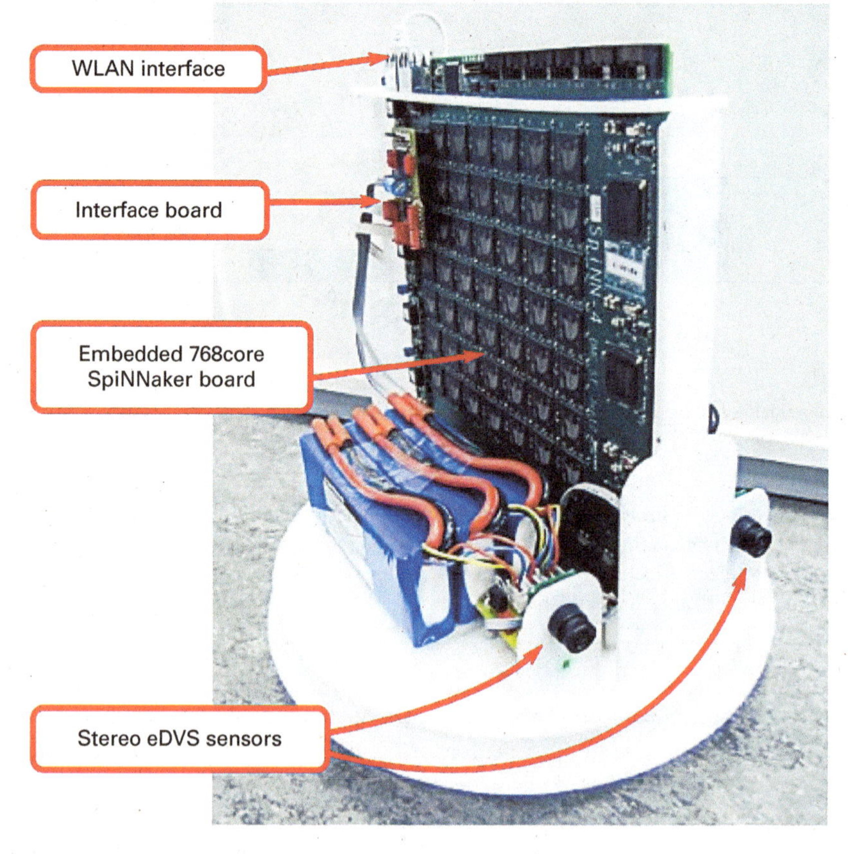

Figure 2.19
Robot platform used in trajectory stabilization experiments. The SpiNNaker neuromorphic chip received visual input from two dynamic vision sensors (eDVS).

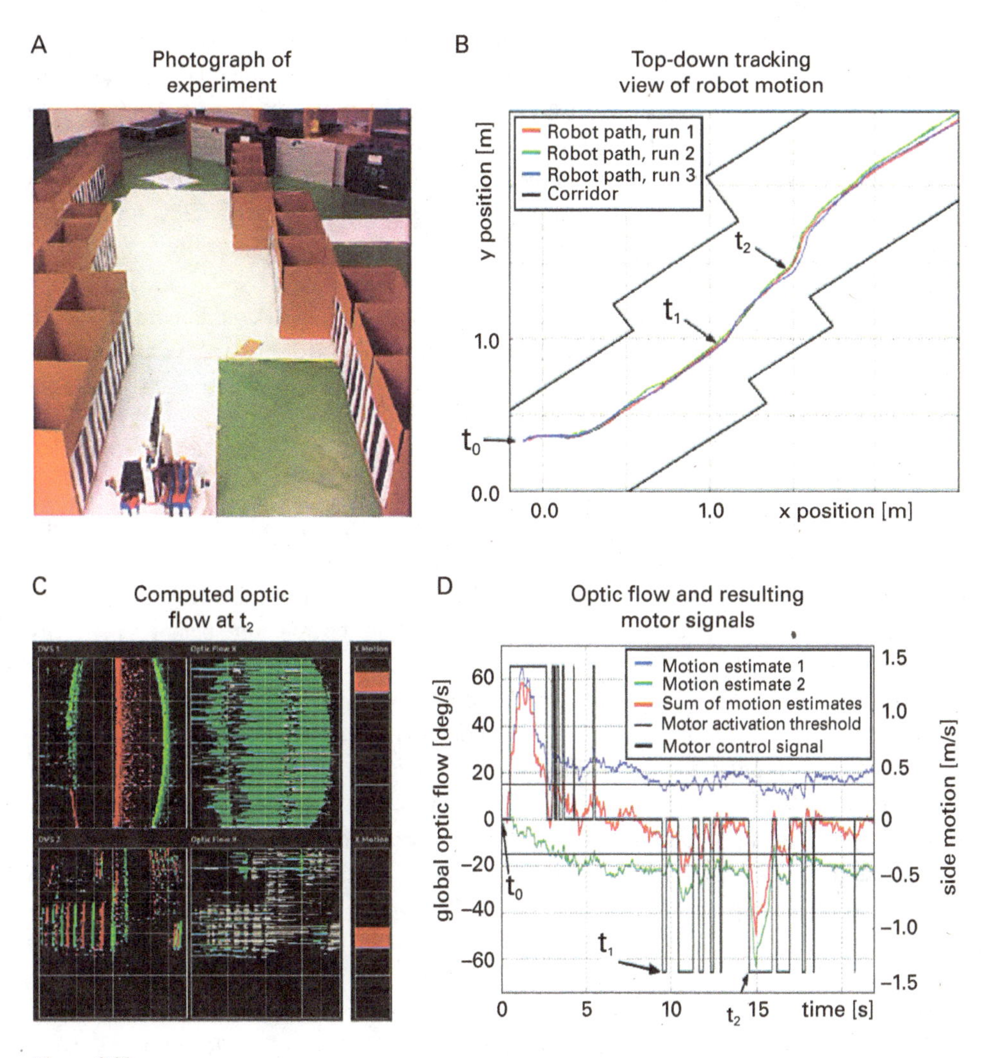

Figure 2.20
Trajectory stabilization using optical flow from eDVS sensors. *A*, Robot traversed a corridor with vertical black and white gratings. *B*, Top-down view of the robot path in three experiments. *C*, Display of observed events from eDVS (*left*). Horizontal local optic flow (green and white indicate different polarities) (*middle*). Combined flow estimates equal to lateral motion motor command (*right*). *D*, Time series of one experiment. The blue and green lines show separate optic flow estimates, the red line shows the global optic flow (sum of the two estimates), and the black lines are the motor command signals.

of figure 2.20*C* (corresponding to the robot's left-facing eDVS) clearly shows a higher spatial frequency than in the lower-left panel (the robot's right-facing eDVS) because of the robot's position close to the left wall (figure 2.20*A*). When the robot was moving all spike trains were processed retinotopically in multiple distributed computing nodes on SpiNNaker. Each node estimated local horizontal OF within a limited region of the total field of view. The resulting horizontal flow vectors are shown in figure 2.20*C* for the left- and right-facing eDVSs, respectively, with green indicating negative flow and white indicating positive (opposite) horizontal flow.

This study showed how a relatively simple bioinspired solution could be implemented in a low-power autonomous system, potentially on a robot or drone with very small batteries. Because the eDVS response time can report pixel changes within a few microseconds, this neurorobotic system could operate at very high speeds.

2.5.2 Cortical Neural Network Model for Visually Guided Robot Navigation

In primates, motion processing is handled primarily by the dorsal stream of the visual cortex (see figure 2.13). In particular, neurons in the middle temporal cortical area (MT) respond preferentially to the speed and direction of visual object motion or self-motion. To investigate whether a cortical model of MT could support a trajectory balancing strategy in cluttered visual scenes, a detailed model of the dorsal stream was embedded on a mobile platform (Beyeler et al., 2015). The architecture of the cortical neural network model is shown in figure 2.21. The model used an efficient GPU implementation of a motion energy model, which is a good approximation for generating cortical representations of motion in the primary visual cortex or V1 (Beyeler et al., 2014; Simoncelli & Heeger, 1998). Spiking neurons in a model of MT then located nearby obstacles by means of motion discontinuities (labeled *Obstacle component* in figure 2.22). The MT motion signals projected to a simulated posterior parietal cortex (PPC), where they interacted with the representation of a goal location (labeled *Goal component* in figure 2.22) to produce motor commands to steer the robot around obstacles toward a goal.

The robot used for the experiments was the body of an R/C racecar with a Samsung Galaxy smartphone mounted on top (see figure 2.21). The robot had IR distance sensors for collision detection. But in these experiments only the smartphone camera was used to detect obstacles. Image frames from the camera were sent via WiFi to a desktop computer with a GPU that simulated the spiking neural network model. Motor commands from the parietal cortex area of the model (PPC in figure 2.21) were sent to the robot's speed controller. Even though the model consisted of 40,000 spiking neurons and roughly 1.7 million synapses, the system was able to handle frame rates of 20 Hz without lags.

Using the balance strategy described previously in the neuromorphic robot and in the equation from the honeybee study, optic flow summed up by MT neuron responses at the

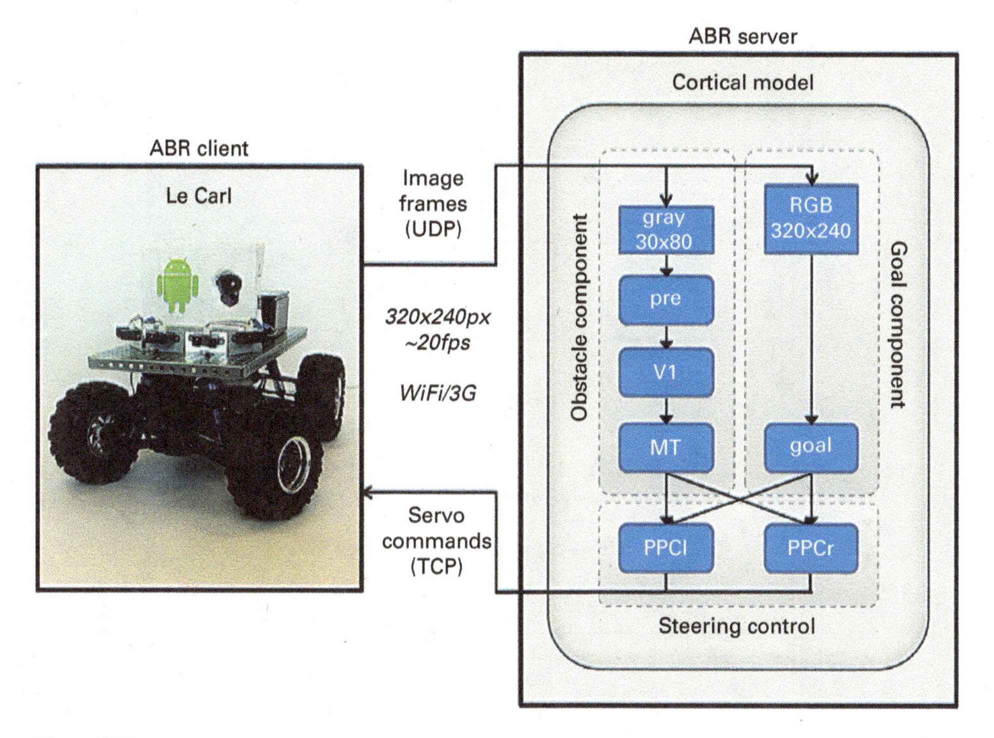

Figure 2.21
Visual navigation using an Android-based robot (ABR) and a model of the dorsal visual cortex. An Android application (ABR client) was used to record 320×240 images at 20 fps and send them to a remote machine (ABR server) hosting a cortical model made of two processing streams: an obstacle component responsible for inferring the relative position and size of nearby obstacles by means of motion discontinuities, and a goal component responsible for inferring the relative position and size of a goal object by means of color blob detection. These streams were then fused in a model of the posterior parietal cortex to generate steering commands that were sent back to the ABR platform. From Beyeler et al. (2015).

PPC region of the model were sufficient to cause the robot to traverse the center of a corridor and avoid obstacles (see figure 2.22). The trajectories were very similar to those observed in human experiments (Fajen & Warren, 2003). Although V1 had directionally selective responses, the robot had smoother trajectories and fewer collisions when the flow signals came from MT. This may be due to the summing and pooling of V1 signals by area MT.

This neurorobot study demonstrated how cortical motion signals in a model of MT might relate to active steering control and suggested that these signals might be sufficient to generate humanlike trajectories. This emergent behavior might not only be difficult to achieve in simulation but also strengthens the claim that MT contributes to these smooth trajectories in natural settings. Even though the neural network was simulated on a

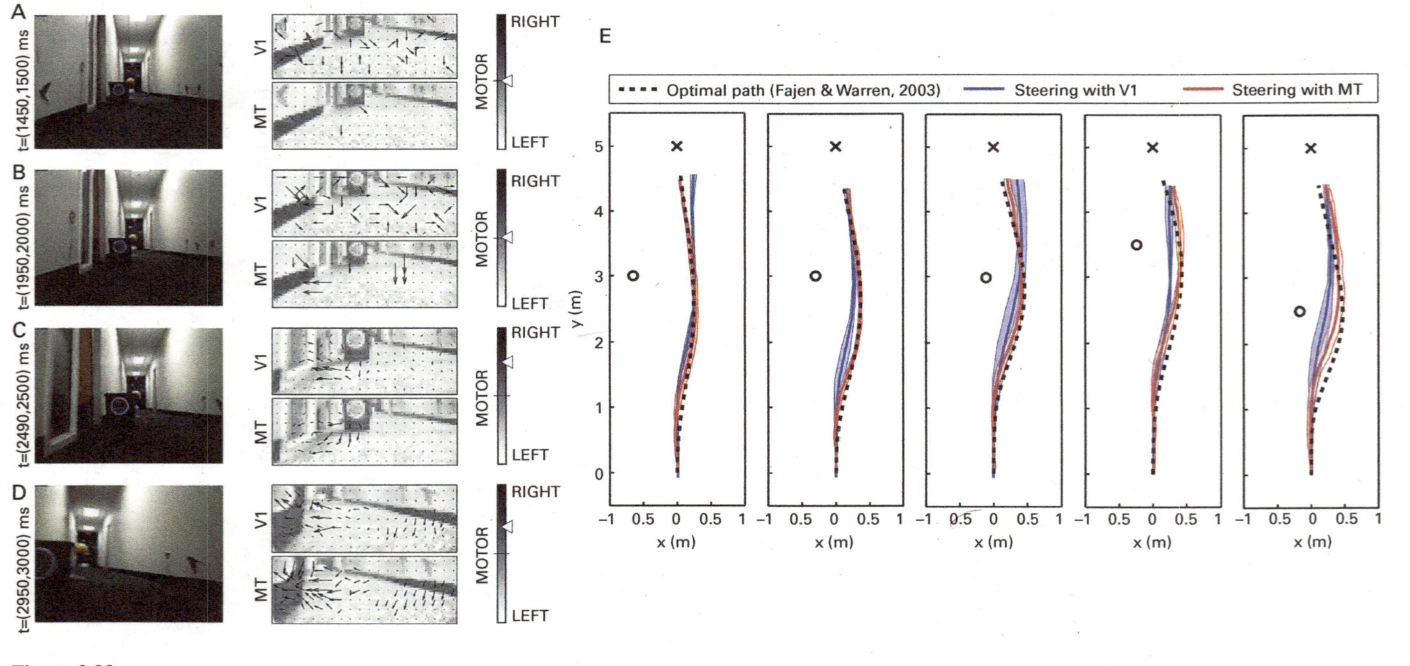

Figure 2.22
A–D, V1 and MT network activity and corresponding behavioral output during obstacle avoidance in a single trial. Each panel summarizes the processing of a single visual input frame, which corresponded to a time interval of 50 ms. *E*, Paths of the robot (colored solid lines) around a single obstacle (recycle bin, represented by **O**) toward a visually salient goal (yellow foam ball, represented by **X**) for five different scene geometries, compared to ground truth (black dashed lines) obtained from the behavioral model (Fajen & Warren, 2003). Results are shown for steering with V1 (blue) as well as for steering with MT (red). Solid lines are the robot's mean path averaged over five trials, and the shaded regions correspond to the standard deviation. From Beyeler et al. (2015).

power-hungry GPU, the spiking neuron model of MT could potentially run on neuromorphic hardware. It would be interesting to replace the conventional camera in this study with the eDVS sensors from the study described in section 2.5.1 to achieve an end-to-end neuromorphic model of the dorsal visual stream.

2.6 Summary and Conclusions

Neuroscience provides the background and inspiration for many neurorobots. It is important to understand systems neuroscience, even at a basic level, to create one's own neurorobots. This understanding makes it easier to read the neurobiological literature for inspiration.

For computer scientists and engineers, neuroscience can seem impenetrable at first. Primarily, it is the terminology that is difficult and arcane. However, anatomy has similarities to circuit diagrams, and many engineering control principles can be applied to brain circuits. The neurons and synapses in the brain can be modeled with the appropriate activation functions and weight updates. However, the complexity of neuroanatomy at multiple levels and the numerous processes to support neurons and systems cannot be completely modeled. Therefore, modelers must choose a level of abstraction on the basis of the scientific questions they want to answer.

For biologists and psychologists, robotics and neural networks can seem daunting at first. But as we saw in the case study, simple brain-inspired algorithms can lead to naturalistic behavior and can scale up to networks with similar architecture as the visual cortex. The next two chapters provide an introduction to the computational tools needed to get started on one's own neurorobot experiments.

2.A Appendix

2.A.1 Neuron Firing Rate

The firing rate, r, of a neuron can be given by:

$$r = \frac{\sum_{t=1}^{T} \delta(t)}{T}, \tag{2.1}$$

where $\delta(t)$ is 1 if the neuron spiked at time t; otherwise it is 0.

2.A.2 Leaky Integrate-and-Fire Neuron (LIF)

The LIF neuron is described by the following equation:

$$V_m(t+1) = V_m(t) + \frac{1}{\tau_m}(E_L - V_m(t) + R_m I(t)),$$
$$\textit{if } V_m > V_{\text{thr}} \textit{ then } V_m = V_{\text{reset}}, \tag{2.2}$$

where V_m is the membrane potential; τ_m is the membrane time constant; E_L is the leak current; R_m is the membrane resistance; I is the synaptic input; V_{thr} is the threshold; and V_{reset} is the membrane potential after a spike. Figure 2.4 shows the response of a LIF neuron with E_L and V_{reset} equal to –65 mV; τ_m equal to 10 ms; R_m equal to 10 MOhm; and V_{thr} equal to –50 mV.

2.A.3 Mean Firing Rate Neuron Models

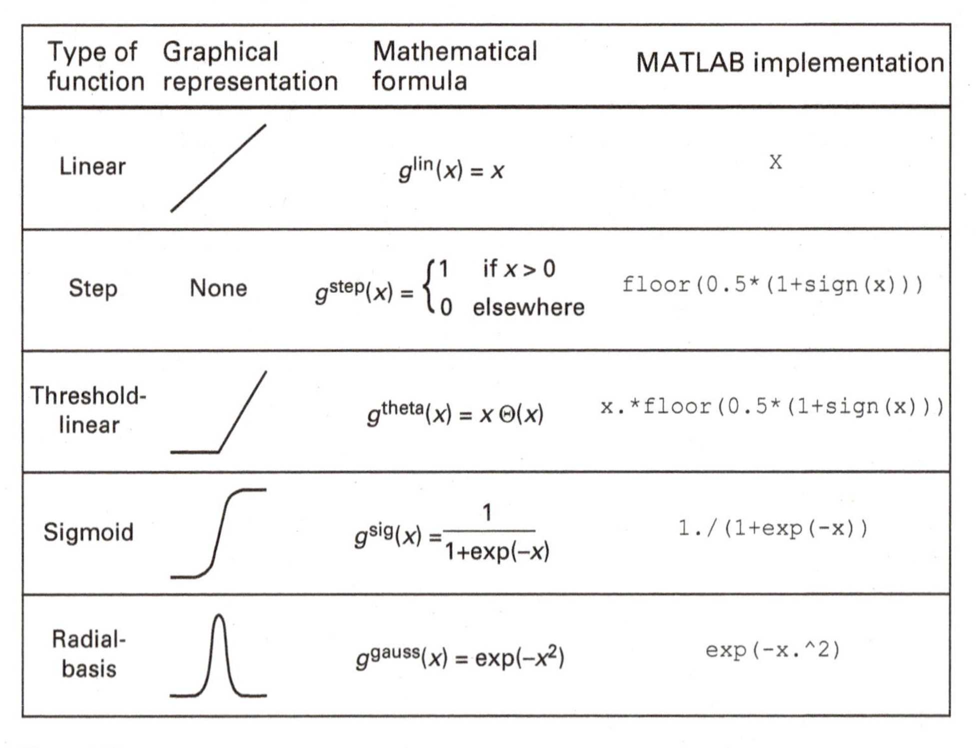

Type of function	Graphical representation	Mathematical formula	MATLAB implementation
Linear		$g^{\text{lin}}(x) = x$	`x`
Step	None	$g^{\text{step}}(x) = \begin{cases} 1 & \text{if } x > 0 \\ 0 & \text{elsewhere} \end{cases}$	`floor(0.5*(1+sign(x)))`
Threshold-linear		$g^{\text{theta}}(x) = x\,\Theta(x)$	`x.*floor(0.5*(1+sign(x)))`
Sigmoid		$g^{\text{sig}}(x) = \frac{1}{1+\exp(-x)}$	`1./(1+exp(-x))`
Radial-basis		$g^{\text{gauss}}(x) = \exp(-x^2)$	`exp(-x.^2)`

Figure 2.23
Adapted from Trappenberg (2010).

2.A.4 Modeling Synapses

Neuron models get their input from other neurons. Therefore, we need a way to calculate the synaptic input into the neuron. For spiking neurons, this would be

$$I_j(t) = \sum_{i=1}^{N_{pre}} w_{ij}\delta(t - t_i), \tag{2.3}$$

where w_{ij} is the weight of a synaptic connection from neuron i to neuron j; and δ is the Dirac function, which is 1 if neuron i spiked at time t and is 0 otherwise. If this were a rate neuron, the equation would change to the following:

$$I_j(t) = \sum_{i=1}^{N_{pre}} w_{ij}n_i(t), \tag{2.4}$$

where n_i is the presynaptic activity of neuron i at time t. Equations (2.3) and (2.4) can be changed to more realistically describe synapse conductance. For conductance-based synapses, interested readers are referred to textbooks on computational neuroscience (Abbott & Dayan, 2005; Miller, 2018). As we discuss in the next chapter, synapses are plastic; that is, they can change over time.

3 Learning and Memory

3.1 Introduction

Learning and memory are two closely related concepts that enable neurorobots to adapt to a dynamic environment. To understand their importance in embodied agents, one can imagine how cognition might have developed in the brain across many years of evolution. Simple organisms without developed brains have existed by having features that enable survival within each organism's niche. Without learning, when environmental change occurs the organism may fail to thrive or may adapt via genetic mutations within the species. In this case, change and adaptation occur slowly over entire generations. The amount of adaptation over the course of an individual's lifetime without learning and memory is highly limited.

With the addition of learning and memory, it becomes possible to adapt much faster within one's lifetime and to learn from past experience. For instance, an organism may learn new associations between stimuli and relearn these associations when the environment dynamically shifts. Furthermore, the associations that the brain learns are not arbitrary but are prioritized by the risks and rewards that they bring. To survive, associations that have to do with resources or danger are usually learned more quickly than associations about neutral observations in the environment. In simpler learning mechanisms, there may be short-term paired associations such as, "I found food here, so I'll come back tomorrow." On the more sophisticated end, a system of learning and value can cover a long time span and very abstract inputs, such as "Good thing I learned how to start a campfire in my youth, as I am lost in the wilderness."

In this chapter, we cover the most common learning rules of biology and artificial intelligence and how to apply them in the building of neurorobotic models and experiments.

3.2 Learning Types

There are three main classes of learning rules used in the modeling of learning systems and neurorobotics: **unsupervised learning**, **supervised learning**, and **reinforcement learning** (figure 3.1). Unsupervised learning refers to the absence of a teacher to tell the learner

A Supervised learning B Unsupervised learning C Reinforcement learning

Figure 3.1
Categories of learning in AI and neurorobotics. *A*, Supervised learning occurs when an external teaching signal is available to guide an agent's behavior to learn a desired task, such as a sports coach. *B*, Unsupervised learning occurs with unstructured observations, such as repetitive movements transitioning from clumsy to smooth. *C*, Reinforcement learning occurs when there is positive feedback (e.g., scoring a goal) or negative feedback (e.g., receiving a penalty) from the environment. The agent can learn which moves led to desired outcomes by exploring the action space.

whether a particular piece of knowledge is correct or not, whereas supervised learning has a teacher or at least some external signal beyond the agent itself to help determine what is good or not. Reinforcement learning occupies a position somewhere between these two classes: a teacher is not present, but the learner receives positive or negative feedback from the learner's actions in the environment. An example of supervised learning is having a sports coach carefully watch your actions and tell you what to improve. An unsupervised learning example is visiting a playing field and gaining familiarity with the equipment and line markers. Reinforcement learning is exemplified by playing a few games of the sport and discovering which actions lead to good outcomes. Each of these types of learning has advantages and disadvantages. Although a coach can quickly improve your performance in a targeted manner, a coach is not always available. Furthermore, discovering things for yourself in a real game can help you develop new insights. If you are not completely aware of the rules of the game, even completely naïve play can build structural knowledge that can be applied later or adapted to a different but overlapping task.

These same considerations are used in modeling learning in neurorobotics experiments. If a specific task is to be taught to a robot, supervised or reinforcement learning works well, but if the task is general, such as recognizing different categories of objects, unsupervised learning may be more useful. Whereas all three types of learning can be observed in the brain to some extent, it is rare to have pure supervised learning. Even if a physical teacher agent is present, the brain still uses some value system to react to the instructions of the coach. Pure supervised learning is explored more in traditional machine learning and is less prevalent in biological learning.

3.3 Neural Network Basics

The majority of learning models we will cover in this book apply to neural network models of cognition. Although there are many possible implementations of neural networks, they all rely on a common basic structure and similar notation. Figure 3.2*A* shows an example of what may be the simplest neural network possible. The neurons of a neural network are interchangeably referred to as units, and the synapses are interchangeably referred to as connections or weights. At a minimum, a neural network should have an input, an output, and a way to tie the output to the input. To represent this, the network shown in figure 3.2*A* has an input unit x_0, an output unit y_0, and a weight w_{00} connecting them. Each of these variables represents a single numerical value. The input typically represents the strength of some stimulus that the agent perceives, and the output can either be the strength of another stimulus or the strength of some action taken by the agent. The weight is the strength of the connection between the input and output, which can be learned through experience. In describing two neurons connected by a weight, the input neuron is the presynaptic neuron and the output neuron is the postsynaptic neuron. The output value is calculated by multiplying the input value by the weight. If there is a strong relationship between the input and output and the input is large, multiplying the input with the weight will yield a high output value.

The simple neural network model shown in figure 3.2*A* is already sufficient for many models of learning. However, for more complex stimuli, it is necessary to extend the network to include multiple units and layers. Figure 3.2*B* expands the network to include multiple

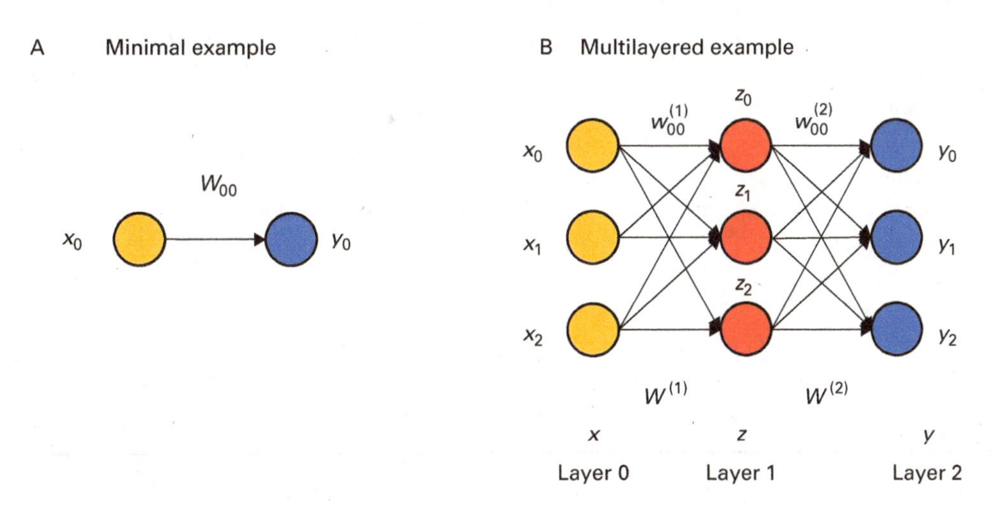

Figure 3.2
Examples of basic neural networks. *A*, Minimal example of a neural network, including an input unit, output unit, and weight. *B*, Multilayered example of a neural network, including an input layer, hidden layer, and output layer. Each layer is represented by a vector consisting of values for each unit in the layer. Weights between each layer are represented by 2D matrices.

inputs and outputs, as well as a hidden layer. The hidden layer can be thought of as some intermediate step of processing performed by some area of the brain. To handle the extra complexity, a little extra notation is needed. Each layer is denoted as a vector in bold text, with each individual unit in the vector referred to by a variable of the layer name and a subscripted index of the neuron. Indices start at 0 and count upward, as is traditional in computer science. The weights between each layer are represented as a two-dimensional (2D) matrix with a superscript of the index of the ending layer of the weights. To access a specific weight value, we use the w variable with the superscript as the postsynaptic layer index and the subscript as the indices of the presynaptic and postsynaptic neuron. Appendix 3.A.1 covers the vector and matrix operations necessary to calculate the activities in a neural network.

Hebbian learning is a fundamental learning type in unsupervised learning (Hebb, 1949). As mentioned in chapter 2, Hebbian learning captures associations between two stimuli. For instance, a fire alarm often consists of both flashing lights and a piercing sound. The brightness of the alarm could be represented by a value in the range of 0 to 1, and the loudness of the alarm could also be represented by a value from 0 to 1. By experiencing fire alarms, one begins to associate the light stimulus with the sound stimulus, such that if only the flashing light were observed, the person would still know that there was a fire alarm. A straightforward way to encode this relationship is by multiplying the values of the stimuli. For instance, if light and sound are both observed as 1, the relationship is a strong value of 1. If the light is 1 and sound is 0.33, the relationship is a weaker value of 0.33.

Figure 3.3*A* shows a schematic of how Hebbian learning might work in a small neural network. Given are two groups of neurons, A and B, with two neurons per group. In the scenario, the first neuron in group A (N_{A1}) is co-active with the first neuron in group B (N_{B1}). This will cause the strength of the connection between these two neurons to increase, as denoted by the thicker arrow. Furthermore, this may cause a decrease in the connection between N_{A2} and N_{B1}. The result of this learning is that the next time N_{A1} is active, there is a better chance that N_{B1} is more active. This is called unsupervised learning because the plasticity is driven by intrinsic co-activity rather than by some extrinsic teacher or supervisory signal. Unsupervised learning can find consistencies in environmental stimuli that are important for the organism. Often, we call this experience-dependent learning because these consistencies are discovered as the agent samples its world. For neurorobots, unsupervised learning can result in learning of categories or classes of information from their own exploration. In cognitive science, this classification is called **perceptual categorization**.

In addition to unsupervised learning, value-based learning or neuromodulated learning is important for developing neurorobotics. As the robot explores its world, it may discover something of value. It does not matter if it is good, bad, or merely different. However, it is important to remember this stimulus or event for future reference. In this case, a modulated signal associated with the value can boost learning in the brain. **Neuromodulators** in the brain have the ability to send signals, through the broad release of neurotransmitters, for important events to be learned and remembered (Krichmar, 2008). Examples of neuromodulators include dopamine, serotonin, noradrenaline, and acetylcholine. Figure 3.3*B*

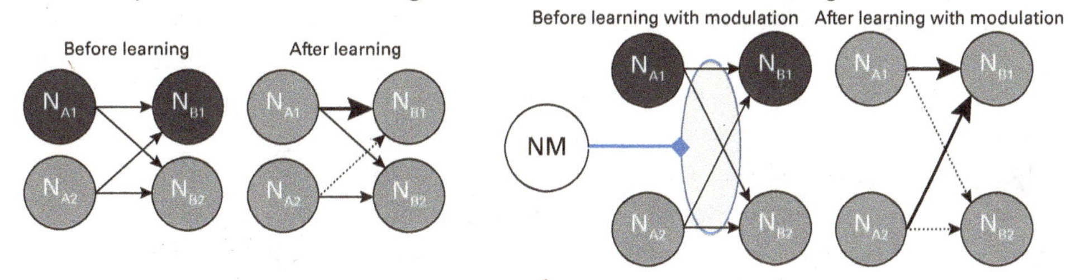

Figure 3.3
Schematic for learning in neural networks. Each circle represents a neuron, and each arrow represents a synaptic connection. The darker the circle, the more active the neuron. The thickness of the arrow denotes the connection strength. In both examples, there are two groups of neurons, A and B, and connections from the A group to the B group. *A*, Unsupervised Hebbian learning. The coactivity between neuron N_{A1} and N_{B1} results in a stronger connection strength between these neurons and a weaker connection between N_{A2} and N_{B1}. *B*, Neuromodulated learning. The activity of neuromodulatory neuron NM signals an important event to be learned. This causes all connections to output neuron N_{B1}, which was active at the time, to be strengthened and connections to N_{B2} to be weakened. Note that after learning the connection strength from N_{A1} to N_{B1} is stronger than N_{A2} to N_{B1} because N_{A1} was more active when the neuromodulatory event occurred.

shows a simplified case of this neuromodulated learning. Neuron NM is a neuromodulator that signals an important event and releases its neurotransmitter to all neurons. The connections between the active neurons are potentiated and the other connections may be depressed. Note that in this example, both connections to output neuron N_{B1} were increased and connections to output neuron N_{B2} were decreased. The effect is that the next time the event occurs or is about to occur, N_{B1} will be active. For example, if the event was seeing a predator, which might be signaled by N_{A1} or N_{A2}, output neuron N_{B1} may trigger an escape response.

Using this intuition, we can understand the equation (3.1) for the Hebbian learning rule. To learn a task, memory is required to persistently store information about what was learned. In the example of the fire alarm, let us encode the light stimulus as the variable x_i and the sound stimulus as x_j. You can think of these as two neurons in the brain that are activated when experiencing these stimuli. Now imagine a connection going from x_i to x_j, which we will call w_{ij}. This is like a synapse between the two neurons. Prior to experiencing anything, the value of w_{ij} is very small because there is no relation between the two. The Hebbian learning rule is expressed by the following equation, which changes the weight value:

$$\Delta w_{ij} = \alpha x_i x_j, \tag{3.1}$$

which is the product of the values of the two neurons multiplied by a learning rate α, which is often between 0 and 1. This means that the weight will increase in proportion to the co-activity of the presynaptic and postsynaptic neurons. The activity of the postsynaptic neuron is

$$x_j = x_i w_{ij}, \tag{3.2}$$

which is updated based on the new weight, w_{ij}, and the activity of the presynaptic neuron, x_i.

In the example of the fire alarm, a single experience will increase the weight value for w_{ij}. The next time the agent experiences a light stimulus like the one from the fire alarm, the sound of the alarm will come to mind, even if the sound is not actually there. Because x_i is active and w_{ij} is high, the signal greatly influences the activity of x_j to be high as well.

Equipped with the Hebbian learning rule, we can apply unsupervised learning to a simple neurorobotics experiment. A robot's task is to approach the brighter side of a room. Figure 3.4*A* shows a schematic depiction of the experiment, with the area split into a red side and a blue side and the light source slightly toward the blue side. A corresponding small neural network encodes the relationship between the light and color stimuli (figure 3.4*B*). There are two input neurons corresponding to the brightness of the blue side and the red side. There are two output neurons corresponding to *move to the left* or *move to the right*. There is a weight connecting the blue intensity to *move to the left* and a weight connecting the red intensity to *move to the right*. Let's work through an example, as follows:

If the initial values of the network are

Learning rate: $\alpha = 0.25$;

Output neurons: $n_{left} = n_{right} = 0.5$;

Input neurons: $n_{blue} = 0.75$; $n_{red} = 0.25$;

Weights $w_{BlueLeft} = w_{RedRight} = 0.5$;

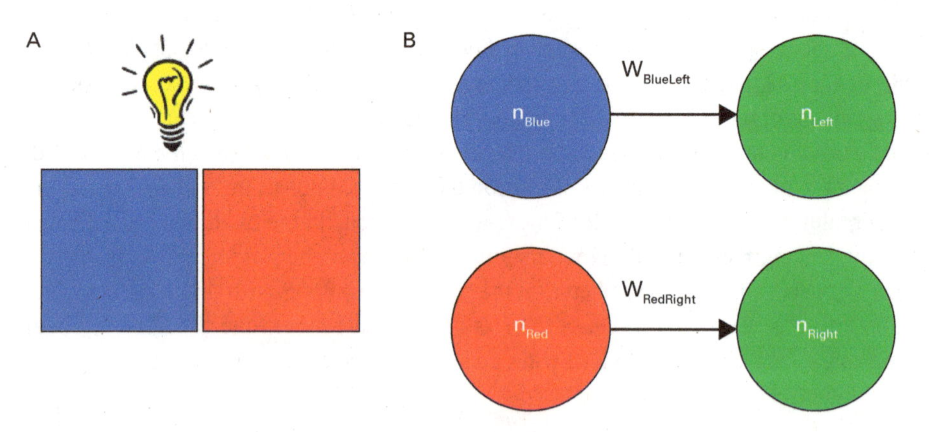

Figure 3.4
A simple neurorobotics experiment involving Hebbian learning. *A*, The task of the neurorobot is to explore an area that is colored blue on one side and red on the other side and to determine which side of the arena is brighter. *B*, The brain of the robot can be represented using a network with two input neurons representing the brightness of the blue and red sides, and two output neurons representing the robot's movement. There are weights between the color and the movement. Note that in this example the light is toward the blue side of the area.

and then if we run the neural network ten iterations following equations (3.1) and (3.2), the result is

$$1: n_{Left}=0.445, w_{BlueLeft}=0.594, n_{Right}=0.133, w_{RedRight}=0.531$$

$$2: n_{Left}=0.508, w_{BlueLeft}=0.677, n_{Right}=0.135, w_{RedRight}=0.54$$

$$3: n_{Left}=0.579, w_{BlueLeft}=0.772, n_{Right}=0.137, w_{RedRight}=0.548$$

$$4: n_{Left}=0.661, w_{BlueLeft}=0.881, n_{Right}=0.139, w_{RedRight}=0.557$$

$$5: n_{Left}=0.754, w_{BlueLeft}=1.01, n_{Right}=0.141, w_{RedRight}=0.565$$

$$6: n_{Left}=0.86, w_{BlueLeft}=1.15, n_{Right}=0.144, w_{RedRight}=0.574$$

$$7: n_{Left}=0.981, w_{BlueLeft}=1.31, n_{Right}=0.146, w_{RedRight}=0.583$$

$$8: n_{Left}=1.12, w_{BlueLeft}=1.49, n_{Right}=0.148, w_{RedRight}=0.592$$

$$9: n_{Left}=1.28, w_{BlueLeft}=1.7, n_{Right}=0.15, w_{RedRight}=0.601$$

$$10: n_{Left}=1.46, w_{BlueLeft}=1.94, n_{Right}=0.153, w_{RedRight}=0.611$$

After these pairings of light and movement, the left neuron, n_{Left}, is much larger than the right neuron, n_{Right}, and the robot should explore the blue side of the room.

So far, the examples of Hebbian learning have been very simple, involving pairwise associations between the perceptions of similar stimuli. They can be extended to include many inputs of different types and multiple levels. The neurons and connections in our Hebbian training example form a simple type of neural network model that can be built upon. The following are three possible avenues for expansion:

1. The number of inputs can be expanded. Rather than only a single input such as the presence of light, one could use a collection of small inputs. Place cells in the brain are one example that can be modeled as the association of many inputs to one output (Derdikman and Moser, 2010). A single place is associated with a set of inputs, such as a collection of landmarks. When these landmarks are activated in the brain, a place cell encoding that particular place activates.
2. The number of layers can be increased. Our Hebbian training example has two layers, an input layer corresponding to the color intensity and an output layer corresponding to the movement direction. We could add another layer of neurons that connects the color to another set of sensors or a motor command to move the robot backward or forward in response to the detection of a color.
3. The learning rule can be modified. There have been many variants of the Hebbian learning rule to improve upon biological plausibility and work within computational constraints.

3.4 Weight Stabilization

A weakness of the Hebbian learning rule is that it has no bounds on the amount of learning that can occur or on the maximum strength of an association. To understand this, let's return to the example of the robot learning to associate the light stimulus with the color blue. At first the weights are small. After each repeated exposure to the pairing and learning using the Hebbian learning rule, both the weights from blue to left and from red to right become larger. At the same time, both output neurons are growing larger. If there were a limit on how large a weight could grow, or if the neuron activity were constrained between 0 and 1 as would be the case in a sigmoid function, eventually both the weights and the output neurons would be saturated.

Weight normalization is one strategy for stabilizing the updating of weights. This strategy involves placing a limit on the weight values of the units, for example, that ensures that the sum of the weights going to a unit never exceeds a defined value. The effect is that each time a weight value is increased, the values of other weights may be decreased. In appendix 3.A.2, we apply weight normalization to the example shown in figure 3.4. This weight normalization scheme is related to synaptic scaling and homeostatic plasticity observed in the brain (Turrigiano, 1999; Turrigiano, 2012).

Another biologically plausible learning rule is the **BCM learning rule**, so called because of the original work by Bienenstock, Cooper, and Munro (Bienenstock et al., 1982; Cooper and Bear, 2012). They found that when neural activity was suppressed by preventing visual input to one eye, the synaptic strength of postsynaptic targets in the visual cortex increased. Likewise, if activity was too high, the synaptic strength of postsynaptic targets decreased. The BCM rule is a modified Hebbian learning rule that has a built-in weight stabilization. Each neuron has a threshold value such that if the postsynaptic neuron has an activity level below the threshold, the weight change is negative or depressed, and if the postsynaptic neuron has an activity level above the threshold, the weight change is positive or potentiated. Depending on the overall activity this threshold can dynamically change to keep weights within an operating region. For example, if activity is low, the threshold decreases and there is more LTP. Conversely, if activity is high the threshold increases and there is more LTD. Appendix 3.A.3 describes the equations for the BCM rule.

Weight stabilization through normalization and the BCM learning rule plays a part in enabling a network to forget information, a skill that goes hand in hand with the skill of learning. Often, the information in an environment becomes outdated or irrelevant. When new information is learned, the old information is weakened naturally through normalization. For instance, an agent may learn to associate a food source with a particular location, but if the food fails to appear for a while or if a new food source appears, the association of food with the old place becomes weaker as the information is no longer needed.

3.5 Classical Conditioning and the Rescorla-Wagner Learning Rule

Classical conditioning is the study of how animals make associations between familiar, neutral stimuli and novel, salient stimuli. Ideas from Hebbian learning still apply, but they are described using a particular formalization of stimulus and response types. The classical conditioning framework consists of a conditioned stimulus, conditioned response, unconditioned stimulus, and unconditioned response. An **unconditioned stimulus** (US) has a preexisting habitual response, which is the **unconditioned response** (UR). A **conditioned stimulus** (CS) is the previously neutral stimulus with which we wish to build a new association. The **conditioned response** (CR) is either the new response that we wish to associate with the CS or is the same as the UR but now is driven by the CS. Classical conditioning is a mechanism by which the US is used to train the CS.

The most famous example of classical conditioning is the series of experiments performed by Ivan Pavlov (Pavlov, 1929). In these experiments, a dog would salivate at the presence of food. The food is the US and the salivation is the UR. For many trials, food was administered along with the sound of a bell, which is the CS. After a while the dog would salivate at the sound of the bell even if the food was not present. In this example, the CR is the salivation at the presence of the bell.

The **Rescorla-Wagner rule** is often used to model classical conditioning (Rescorla and Wagner, 1972). Rescorla and Wagner proposed a mathematical model to explain the amount of learning that occurs on each trial of Pavlovian learning (Wilson, 2010). They assumed that more learning occurs when there is surprise, that is, when the feedback given a stimulus does not match one's expectation. This implies that the agent is trying to learn which stimuli predicts an outcome. Compared to Hebbian learning, which implies that any two co-occurring stimuli become associated with each other, the Rescorla-Wagner rule accounts for observations in behavioral research that show a more complicated story. Given the richness of stimuli in natural environments, it is impossible to learn paired associations for everything. The context of what other stimuli are present greatly influences whether any particular association is made. The equation for the Rescorla-Wagner rule is

$$\Delta V = \alpha\beta(\delta - \Sigma V), \tag{3.3}$$

where ΔV is the change in expected value; α is the salience of the CS; β is the learning rate; δ is the outcome; and the summation of V is the expected value. Following equation (3.3), if we set α to 1.0, β to 0.10, and a reward payoff ($\delta = 1.0$) for the first fifty trials, we see the expected value, V, reach the actual value of δ by trial 50 (see blue markers in figure 3.5). Now if we stop delivering a reward after trial 50, the expected value drops to 0 by trial 100. The red markers in figure 3.5 show the expected value when the payoff rate is 50 percent for the first 50 trials. The drop in expected value for the second fifty trials when there is no payoff is known in experimental psychology as **extinction**.

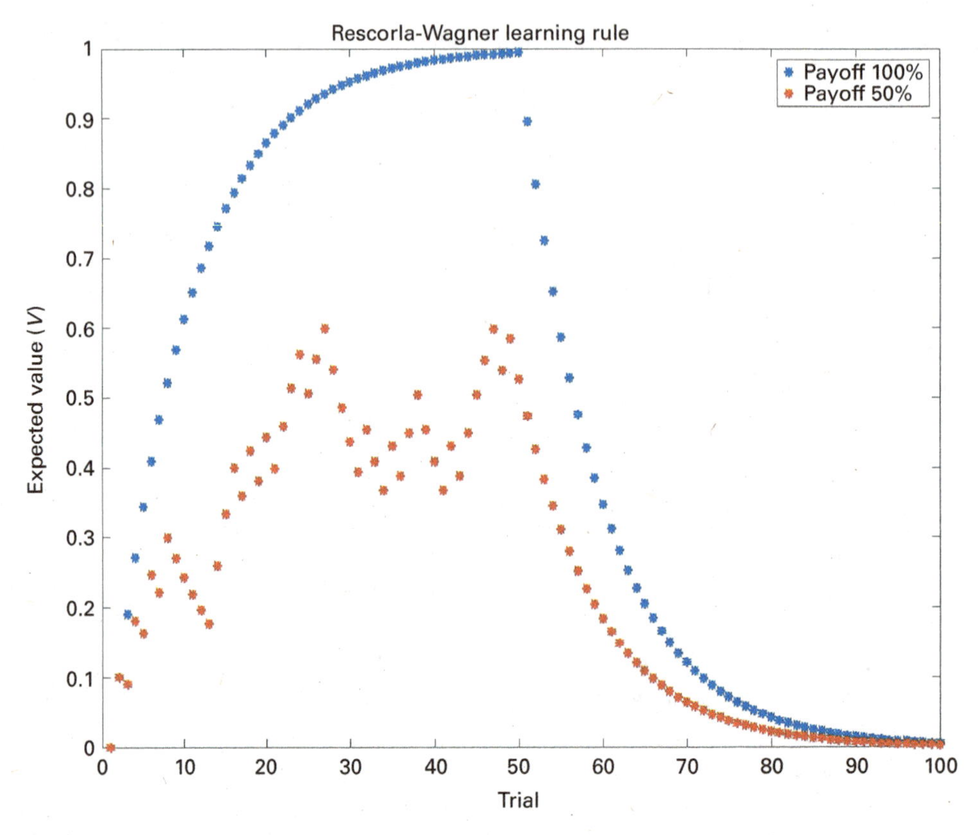

Figure 3.5
Rescorla-Wagner learning rule. The plot shows the expected value, V, or 100 trials. The salience, α, is set to 1.0; the learning rate, β, is set to 0.10; and the reward or outcome, δ, is set to 1.0. The reward is delivered for the first fifty trials at a rate of 100 percent (blue markers) and 50 percent (red markers). After trial 50, the reward is no longer delivered.

The Rescorla-Wagner rule can account for many behavioral phenomena, such as the effect known as blocking. This occurs when the CS is already strongly paired with other stimuli and the CS-US association is blocked. As an example of blocking, let us return to the robotics experiment of the red and blue stimuli. Suppose, as shown in figure 3.4, that we want to condition the robot to drive toward blue when it already drives toward light. V is the expected value that blue predicts the light. If the light and the blue color are the only stimuli in the experiment, learning the association is not a problem. However, if there is another stimulus, such as red, with which the robot has made prior associations and perhaps has already been conditioned to drive toward, this may prevent learning of the light and blue pairing. The reason is that V would already have a high value due to the association value of red being high and thus being a value very close to δ and making the change in association strength of blue very low in applying the learning equations.

To try out the Rescorla-Wagner in a Webots simulation, follow the instructions in box 3.1.

Box 3.1

A Webots simulation that uses Rescorla-Wagner learning to condition a robot's behavior can be found in the RescorlaWagner_Conditioning folder on GitHub: https://github.com/jkrichma/NeurorobotExamples/.

During the simulation, the robot spins in place and stops to stare at different colored objects (figure 3.6). In the first fifty views, the robot gets a reward 90 percent of the time when it looks at red, 25 percent of the time when it looks at green, and 50 percent of the time when it looks at blue. In the second fifty views, the rewards are changed (10% for red, 100% for green, and 50% for blue).

Figure 3.6
Robot for Rescorla-Wagner simulation. A camera is enabled, and different color blocks are placed in the world. The robot spins in place and stares at different colors. The staring length is dependent on the Rescorla-Wagner learning rule and the values associated with each color.

The amount of time the robot stares at each color is proportional to the value associated with that color. For example, figure 3.7 shows the amount of time that the robot looked at red over one hundred trials. Note how the amount of time staring at red increases and then rapidly decreases or extinguishes when the reward frequency drops.

(continued)

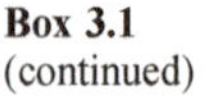

Box 3.1
(continued)

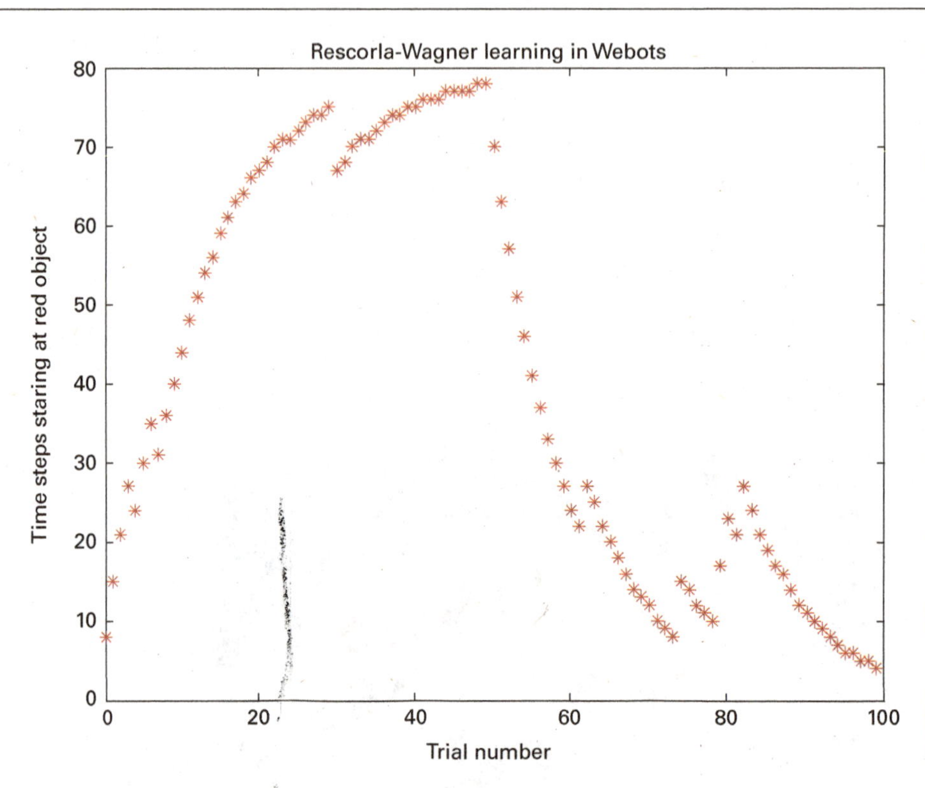

Figure 3.7
Rescorla-Wagner learning rule applied to the robot looking at different objects. The chart shows the amount of time the robot stares at the red object. For the first fifty trials, the robot received a reward 90 percent of the time. For the second fifty trials, it received a reward 10 percent of the time. The reward (δ) was set to 2.0 in the simulation.

Plot the staring time for the other colors. Change the reward delivery and observe how that affects behavior. How does changing the value (i.e., δ) or the learning rate affect conditioning?

3.6 Learning and Memory in Spiking Neural Networks

As we explored in chapter 2, spiking neural networks are a special type of neural network that examines the spiking activity of each neuron in the brain. As opposed to traditional neural networks, which represent the activity of a neuron as a continuous value, spiking neurons fire a discrete action potential. The timing of this binary event is important for learning and leads to a learning rule known as spike-timing-dependent plasticity (figure 3.8).

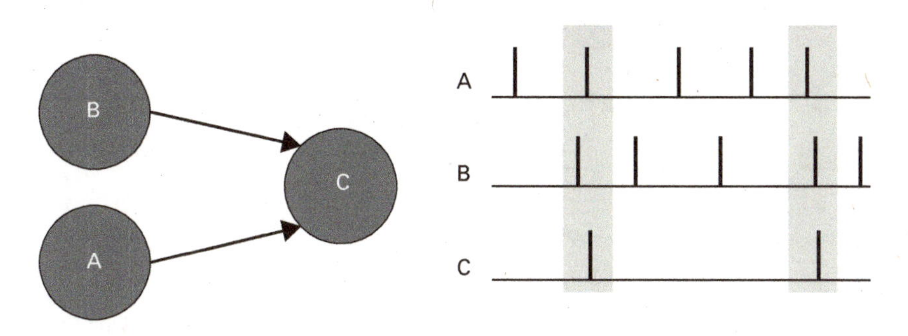

Figure 3.8
Neuron signals involved in spiking neural networks. Neuron C integrates signals from neurons A and B. If enough incoming spikes occur in a short span of time, neuron C spikes. Sample spike trains are shown on the right. Picture adapted from www.wikipedia.com.

3.6.1 Spike-Timing-Dependent Plasticity

Spike-timing-dependent plasticity (STDP) is similar to Hebbian learning in that learning is related to the correlation between presynaptic and postsynaptic activity, but it also considers the fine-scale timing of spiking neurons (Song et al., 2000). This makes it an appropriate learning rule for spiking neural networks. As Hebbian learning changes connections according to their co-activity, STDP changes connections according to how closely in time two neurons spike. In the typical STDP formulation, if the presynaptic neuron fires shortly before the postsynaptic neuron, the connection is strengthened. If the opposite ordering occurs, the connection is weakened. STDP is a versatile rule that allows one to define different weight-changing behaviors on the basis of the spacing and order of the spikes of the presynaptic and postsynaptic neuron. The STDP rule is expressed by the following equations:

$$W(x) = A_+ \, exp(-(t_{\text{post}} - t_{\text{pre}}) / \tau_+) \text{ for } (t_{\text{post}} - t_{\text{pre}}) > 0 \quad (3.4)$$

$$W(x) = -A_- \, exp(-(t_{\text{post}} - t_{\text{pre}}) / \tau_-) \text{ for } (t_{\text{post}} - t_{\text{pre}}) < 0 \quad (3.5)$$

where t_{post} is the time of the postsynaptic spike, t_{pre} is the time of the presynaptic spike, A_+ and A_- are the amplitude of potentiation and of depression, respectively, and τ_+ and τ_- are time constants to how much the amplitude decays as the time between presynaptic and postsynaptic spikes increase. The equation handles the weight update differently on the basis of whether the postsynaptic or presynaptic neuron spiked first. Figure 3.9*A* shows a few examples of different parameters set for the weight change. The top left subplot of figure 3.9*A* shows what the STDP function would look like if A_+ and A_- were set to 1 and τ_+ and τ_- had the same value. If neuron *i* spikes before neuron *j*, the amount of weight change is positive. On the other hand, if neuron *i* spikes after neuron *j*, the weight change is negative. The closer together in time the spikes occur, the larger the changes in weight, with the

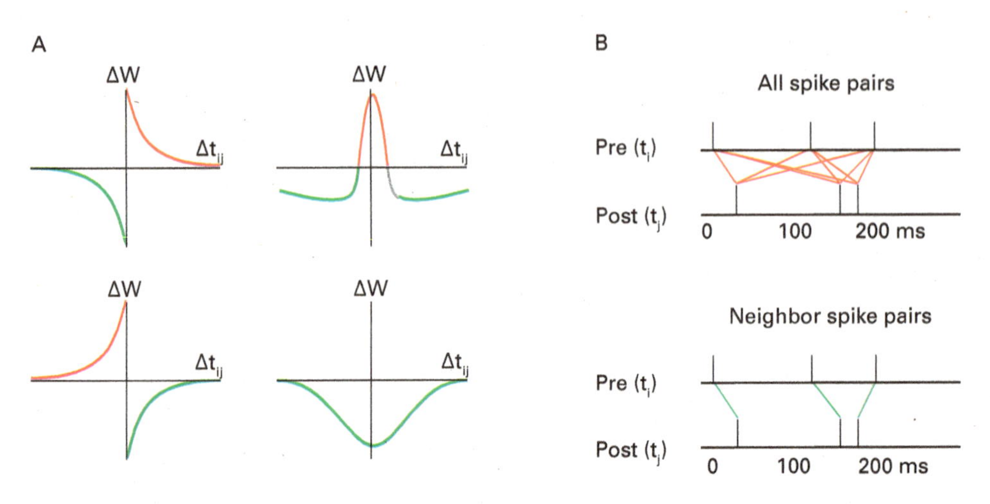

Figure 3.9
Illustration of spike timing dependent plasticity (STDP). *A*, Examples of weight updating function with different parameters. Red lines indicate an increase in weight and green indicates a decrease. *B*, Two methods of comparing spike time pairs. The top shows STDP applied over the full time period with all pairwise spikes compared. The bottom shows only neighboring spikes compared. Picture adapted from Shouval et al. (2010).

amount of change decreasing exponentially as the time between the spikes increases. This means that if the spikes are very far apart in time, they have little effect on each other.

The span of time over which the spike trains are compared can be adjusted in different ways as well (Shouval et al., 2010). Figure 3.9*B* shows two different ways of comparing the spike trains. The first way is to do a complete pairwise comparison of all possible pairs of spikes between both spike trains in the entire period, which is the additive version of STDP. This allows neurons to affect each other over longer, overlapping ranges of time for interesting dynamics. However, one must remember to add asymmetry to the STDP function, because if it is completely symmetrical and each pair is compared twice but with swapped orders, the weight changes will negate themselves and no overall weight changes will occur. For instance, setting all STDP parameters to 1 will not cause any learning to occur if all possible spike pairs are compared. Another way is to consider only the most recent spike; this kind of comparison is called nearest neighbor STDP.

In neurorobotics, STDP would be applied if a spiking neural network is used to model the neural processing of the robot or if special neuromorphic hardware is used to run the robot. There are also cases in which the higher level of biological detail of the spiking model of learning would be useful. A spiking neural network may be desirable if precise timing is required in the neurorobotic experiment. In this case, one might want to implement STDP for the learning. Models involving temporal information, such as auditory localization for robotics, could be modeled very efficiently using spike trains and STDP to help a robot get around using auditory input.

3.7 Summary and Conclusions

Learning and memory are essential for agents to be successful in dynamic environments. There are three major types of learning: unsupervised learning, supervised learning, and reinforcement learning. The biological mechanisms behind these types of learning may be modeled by implementing learning rules that capture broad observations of how animals learn. Hebbian learning shows how associations between stimuli may be made, which is perhaps the most elementary kind of learning that can occur. The Rescorla-Wagner learning rule goes deeper into learning observations, showing how contextual stimuli, previous knowledge, and salience can greatly impact the ability to learn a particular association. The association of two stimuli is tied to the actions of the agent to allow the agent to maximize success. These learning rules may also be implemented in spiking neural networks using STDP. In chapter 4, we discuss how value systems are further used to decide what associations are made, how well they are made, and how these effects are modeled using reinforcement learning.

3.A Appendix

3.A.1 Vector and Matrix Operations

To understand how activity is sent through a multilayered network, it is necessary to understand a few concepts from linear algebra. This section gives a brief review of the essential concepts. A vector is a collection of single numbers arranged in a row or a column. The activity of each layer in a network is represented by a vector. A matrix is a 2D grid of single values consisting of rows and columns. A matrix can be formed by stacking multiple columns of vectors horizontally or vertically. Matrix multiplication combines the values of two matrices by finding the products between every row in the first matrix and every column in the second matrix. A *dot product* multiplies vector elements by a column of matrix entries:

$$\boldsymbol{n} = [n_0 \; n_1 \; n_2] = [0.2 \; 0.9 \; 0.4]$$

$$\boldsymbol{W} = \begin{bmatrix} w_{00} & w_{01} & w_{02} \\ w_{10} & w_{11} & w_{12} \\ w_{20} & w_{21} & w_{22} \end{bmatrix} = \begin{bmatrix} 0 & 1 & 0 \\ 0 & 0 & 3 \\ 2 & 1 & 0 \end{bmatrix}$$

Figure 3.10 specifies a three-neuron network. If we were to calculate the activity of neurons, we would use the dot product given by equation (3.6) to get the synaptic input:

$$\boldsymbol{I} = \boldsymbol{n} \cdot \boldsymbol{W} = \sum_0^{N_j} I_j = \sum_{i=0}^{N_i} \sum_{j=0}^{N_j} n_i w_{ij}, \tag{3.6}$$

where $\boldsymbol{I}$ is the synaptic input; i is the index to the presynaptic neuron or the column of matrix $\boldsymbol{W}$ shown in figure 3.3; j is the index to the postsynaptic neuron or the column of matrix

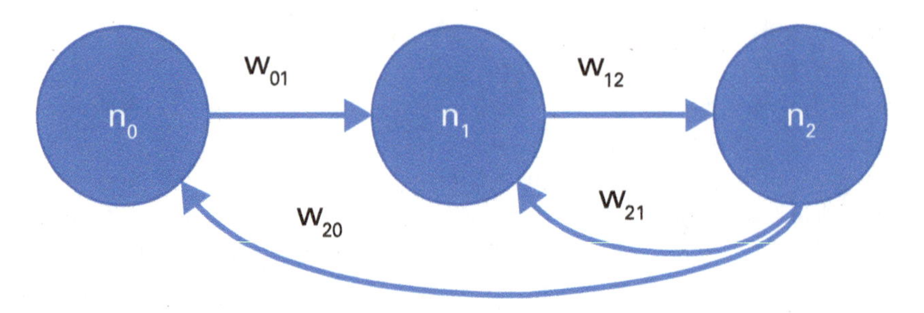

Figure 3.10
Neural network specification. The vector, ***n***, represents the three neurons in a network. The matrix, ***W***, represents the weights between neurons where w_{ij} denotes a connection weight from n_i to n_j. A nonzero number denotes a connection, and the value corresponds to the weight. Zero-valued entries mean that there is no connection between these neurons. The schematic at the bottom shows the resulting network architecture.

W shown in figure 3.10; N_i and N_j is the upper index of the presynaptic and postsynaptic neurons, respectively, which in our example is 2. Remember that programmers typically count from 0. This would yield the following synaptic inputs for each of our neurons:

$$I_0 = n_0 w_{00} + n_1 w_{10} + n_2 w_{20} = 0.2*0 + 0.9*0 + 0.4*2 = 0.8$$

$$I_1 = n_0 w_{10} + n_1 w_{11} + n_2 w_{12} = 0.2*1 + 0.9*0 + 0.4*1 = 0.6$$

$$I_2 = n_0 w_{20} + n_1 w_{21} + n_2 w_{22} = 0.2*0 + 0.9*3 + 0.4*0 = 2.7$$

To get the neural activity, the synaptic input ***I*** would be the input into any of the activation functions shown in appendix 2.A.3. If we applied the sigmoid function to *I*, the new values of *n* would be

$$n_j = \frac{1}{(1 + e^{-gI_j})}, \boldsymbol{n} = [0.69 \;\; 0.65 \;\; 0.94], \tag{3.7}$$

where n_j is the activity of neuron *j*, *g* controls the slope of the sigmoid functions in which higher values of *g* result in a sharper slope of the S-shaped curve, and I_j is the synaptic input for neuron *j*. In this example, *g* is equal to 1, the resulting activation for the three neurons is given by the vector ***n***.

3.A.2 Weight Normalization Using the Sum of Squares

Weight normalization can be applied after each weight update by dividing every individual weight by the square root of the sum the squares,

$$w_i = \frac{w_i}{\sqrt{\sum_{j=0}^{N_w} w_j^2}}, \tag{3.8}$$

where if w_i increases due to a weight update, all other weights will decrease.

Let's run the example from section 3.3 again, this time applying weight normalization after each weight update. The result over ten iterations would be

1: $n_{Left}=0.722$, $w_{BlueLeft}=0.963$, $n_{Right}=0.0675$, $w_{RedRight}=0.27$

2: $n_{Left}=0.728$, $w_{BlueLeft}=0.97$, $n_{Right}=0.0605$, $w_{RedRight}=0.242$

3: $n_{Left}=0.732$, $w_{BlueLeft}=0.976$, $n_{Right}=0.0542$, $w_{RedRight}=0.217$

4: $n_{Left}=0.736$, $w_{BlueLeft}=0.981$, $n_{Right}=0.0485$, $w_{RedRight}=0.194$

5: $n_{Left}=0.739$, $w_{BlueLeft}=0.985$, $n_{Right}=0.0434$, $w_{RedRight}=0.174$

6: $n_{Left}=0.741$, $w_{BlueLeft}=0.988$, $n_{Right}=0.0387$, $w_{RedRight}=0.155$

7: $n_{Left}=0.743$, $w_{BlueLeft}=0.99$, $n_{Right}=0.0346$, $w_{RedRight}=0.138$

8: $n_{Left}=0.744$, $w_{BlueLeft}=0.992$, $n_{Right}=0.0309$, $w_{RedRight}=0.123$

9: $n_{Left}=0.745$, $w_{BlueLeft}=0.994$, $n_{Right}=0.0275$, $w_{RedRight}=0.11$

10: $n_{Left}=0.746$, $w_{BlueLeft}=0.995$, $n_{Right}=0.0245$, $w_{RedRight}=0.0981$

Note that the weights stay in range. As the weight from blue to left increases, the weight from red to right decreases. It also keeps the neural activity in check.

3.A.3 The BCM Rule

The BCM rule is a Hebbian learning rule with a sliding threshold. The role of the sliding threshold is to keep the weights stabilized. Given a presynaptic neuron, x_i, and a postsynaptic neuron, x_j, the equation to describe the weight change is

$$\Delta w_{ij} = \alpha x_i x_j (x_j - \theta_j), \tag{3.9}$$

where θ_j is the threshold value of neuron j and α is the learning rate.

If the threshold for the postsynaptic neuron is a fixed value, the network activity will not be stable. As an exercise to understand this, you may try to run several iterations of the regular Hebbian learning rule for a single weight with an input and output unit. The weight will be seen to grow exponentially. With a constant threshold value, the $x_j - \theta_j$ portion of the equation will soon be eclipsed by the exponentially increasing weight and x_j value. This can be addressed by having θ_j work as a sliding threshold

$$\Delta \theta_j = \beta(x_j^2 - \theta_j), \tag{3.10}$$

where β is the rate of change for the threshold. To ensure that the threshold does not change too fast, β is usually smaller than α. This updating of the threshold in relation to an exponential term of the postsynaptic neuron causes it to change at a higher rate, which makes the weight values stable. The sliding threshold is similar to weight normalization in that the increase of one weight will prevent the other weights connecting to the postsynaptic neuron from increasing. This occurs because an increase of one presynaptic neuron's ability to cause the postsynaptic neuron may cause the postsynaptic threshold, θ_j, to increase due to increased activity, making it difficult for the other presynaptic neurons to drive post-synaptic activity above the threshold.

4 Reinforcement Learning and Prediction

4.1 Introduction

We have examined how learning can occur in robots that actively sense the environment without receiving rewards or seeking a goal. For animals and artificial agents, this learning becomes useful when it is put into action. In experimental psychology, it is called **latent learning**. Furthermore, although unsupervised learning is certainly an important capability, it is not possible to learn everything in this way. Given the limited amount of time and energy of an animal or robot, choices must be made about what information to encode. These choices are likely to be those that lead to the survival and benefit of the agent on the basis of a **value system**, which may have evolved on the basis of either the ecological niche or a cultural trait of the agent. Simple values, such as the desire for food and shelter, lead to visibly understandable choices in what information to encode, whereas sophisticated value systems such as moral codes lead to less apparent choices that play out over longer time horizons. Rewards and value systems shape how agents explore the environment, which in turn shape how the agents learn. In this chapter, we first show how value functions can shape behavior. We then cover the structure of Markov decision processes, which formalize the way agents make decisions in an environment. From this foundation we cover model-free and model-based reinforcement learning as well as measurements of predictability and action selection.

4.2 Braitenberg Vehicle 4

In previous chapters we explored Braitenberg vehicles 2 and 3, observing how simple wiring between sensors and motors can lead to the emergence of lifelike behavior. The vehicles were either attracted or repelled by the light source, thereby demonstrating two opposing value systems: one that positively values light and one that is aversive to light. These examples of value are simple, such that the more of the desired source, the better, or the less of the undesired source, the better. Braitenberg vehicle 4 expands the role of value, showing that a varied value function results in intricate behaviors. For instance, an

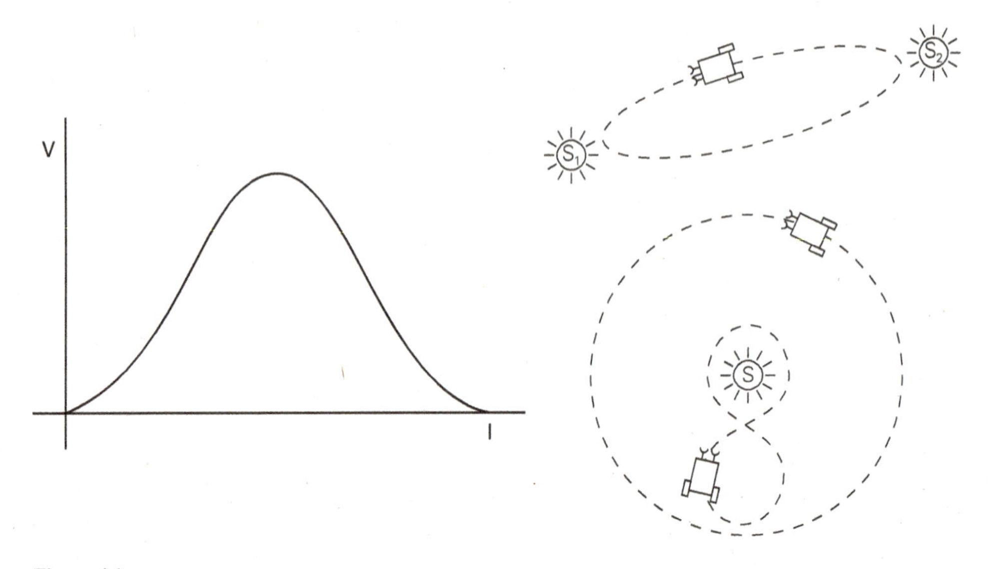

Figure 4.1
Schematic of Braitenberg vehicle 4. *Left*, Example of a varied value function that can lead to interesting emergent behaviors. *Right*, Interesting behaviors that can arise from varied value functions such as circling a light source or oscillating between two sources.

agent could prefer a specific amount of light, striving to stay within the desired range. Figure 4.1 shows a value function illustrating this, in which the highest value is found at a specific intensity of light. If the value is tied to the speed of the motors and two light sources are present, the vehicle may oscillate between the sources to maintain the desired amount of light. If only one source of light is present, the vehicle may circle the light source or perform a figure-eight maneuver around it. With only one value function and a particular environmental setup, an infinite set of trajectories is possible. One can imagine the myriad effects that multiple value sources, environmental factors, and social factors could have on agent behavior. This is the subject of later chapters.

4.3 Markov Decision Processes

The value function controlling the behavior of Braitenberg vehicle 4, which was the relationship between the motor actions and stimulus intensity, was set in advance with no learning required. In many animals, matching value to behavior is not as direct. This leads to the question of how learning mechanisms use value to acquire information. The **Markov decision process** (MDP) formalizes the problem of using reward to learn the best actions to take for a particular environmental state. Each of the terms involved in the MDP has a unique definition, as follows:

1. At any one time, an agent exists within a single **state**, which is a unique descriptor of the environmental conditions relevant to the agent and task. In simple environments such as game boards, the state is often just the location of the agents in the environment. In higher dimensional environments, a state could be highly detailed, such as the pixels from the camera of a robot. The state representation for a task should be carefully selected, with the minimum amount of complexity necessary to capture the desired effects.

2. When the agent is in a given state, a defined set of **actions** is available. The agent chooses a single action out of the set, which brings the agent to another state. For instance, in a game board, an action might be to move from one square to another. For a robot with visual input as state, an action may be to steer left or right.

3. When an action is taken, the next state is not always certain, as some aspects of the environment may not explicitly be part of the model or may not be very predictable. To capture the randomness of the environment, a *transition* from one state to another may be probabilistic. In a game board, choice of an action to move one square may bring the agent forward 90 percent of the time and backward 10 percent of the time (perhaps by a random game element like a dice roll or the move of an opponent). For a robot choosing to steer left to avoid an obstacle, the next state will most likely no longer have the obstacle in view, but there is a chance that the obstacle will move to stay in the field of view, the robot wheels will lose traction, or another obstacle will appear. Through much exploration of the environment over time, these probabilities can be learned and predicted.

4. **Rewards** are typically associated with specific states and drive the agent to learn actions that maximize their accumulated rewards. Rewards are expressed as numeric values and can be positive or negative. They may be sparse throughout the environment, such as a singular large number for winning or losing a game, or they may be interspersed as small rewards throughout the environment. A robot navigating around obstacles may encode hitting the obstacles as a penalty. It is more challenging to learn good actions in an environment with sparse rewards, as it may take a large amount of random exploration to reach a rewarding state. If the agent can learn which states are rewarding, it can then use its knowledge of state transitions to pick actions that it thinks will eventually lead to positive rewards.

5. The **policy** of an agent describes an action plan that an agent uses to pick actions. Before exploring the environment, the policy is likely to lead to random behaviors. Over time, as the agent learns transitions and rewards, the policy improves towards becoming an optimal policy. A policy can be described as a mapping from states to actions. In the case of obstacle avoidance, the optimal policy would likely result in action choices that steer the robot in the opposite direction of the perceived obstacles.

In the MDP, the agent is always within a single state and uses only information from the current state to decide what action to take. The agent performs this action and potentially

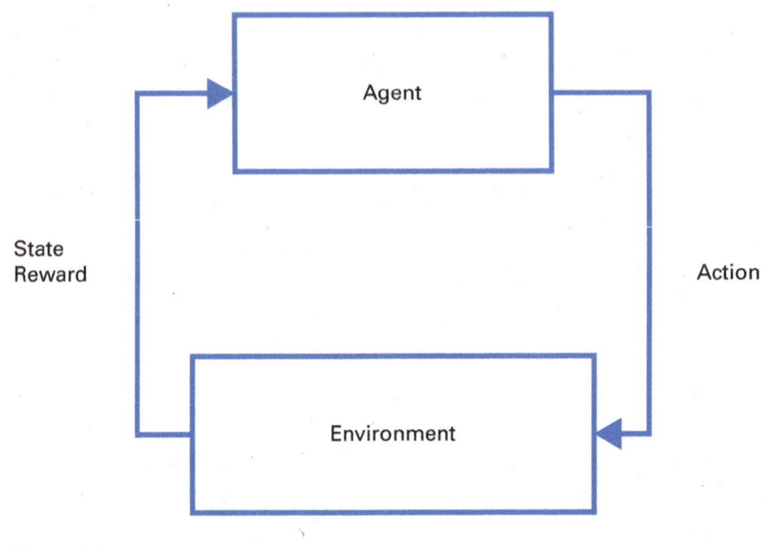

Figure 4.2
The Markov decision process for reinforcement learning.

receives a reward from the environment. The agent is then in a new state, and the process repeats until some termination state is reached (see figure 4.2).

4.4 Reinforcement Learning

MDPs provide a framework for formalizing problems in reinforcement learning, which is the process of learning how to maximize rewards while performing a task in an environment. Whereas learning may be facilitated by an explicit supervisor or teacher, reinforcement learning covers the case of agents exploring environments to discover which actions lead to high value. Reinforcement learning can be split into model-free and model-based categories. **Model-free reinforcement learning** does not create a model of the environment; that is, it does not learn the transition probabilities of state transitions. Rather, its policy is to choose the best action for a given state. Policies learned under model-free reinforcement learning represent which states are good or bad but cannot simulate or explain why. For instance, a robot may learn that it is generally good to steer away from objects, without knowing the model of how the steering moves the robot to a safer location. On the other hand, **model-based reinforcement learning** explicitly creates a model of the environment, which can be in the form of transition probabilities between states. Although this may ultimately lead to a better policy and explanation of actions, the cost of learning and storing transition values presents a trade-off dependent on environmental complexity. One determining factor for using model-based reinforcement learning is whether the model representation would be useful for something else. For instance, if the reward function ever changes unexpectedly, the agent does not have to re-explore the environment as much as before, as it already knows the state transitions. Interestingly, there is evidence for both model-free and model-based reinforcement learning in the brain.

4.4.1 Model-Free Reinforcement Learning

A common algorithm for reinforcement learning is *Q-learning*. The *Q value* is a measurement that relates to the quality of an action, that is, the cumulative amount of reward if the given action is taken in the given state. For every possible state and action pair, there exists a Q value. At first, the Q values may be initialized as 0 or random values. As the agent explores the environment, its state changes with every action at each time step. For each state transition from time step t to $t+1$, the Q value for each state-action pair is updated using the following rule:

$$Q^{new}(s_t, a_t) = Q(s_t, a_t) + \alpha(r_t + \gamma \max_a Q(s_{t+1}, a) - Q(s_t, a_t)). \quad (4.1)$$

The terms of the equation are defined as follows: Q^{new} is the updated Q value for the state-action pair at the current time s_t is the state at time t, a_t is the selected action at time t, α is the learning rate (usually between 0 and 1), r_t is the reward value received at time t, and $\max_a Q(s_{t+1}, a)$ is the maximum value that could be reached in the next state considering each possible action a. The value of the next state is discounted by γ because it may be undesirable to wait too long for a future reward.

The new Q value is the sum of two parts, which are weighted by learning rate a. This learning rate determines how quickly the agent updates its reward information when given a new observation. The first value in the sum is the amount of reward received at the current state. The second value is known as the temporal difference, which is roughly the amount of Q value gain that the agent receives from transitioning from the current state to the next state. The temporal difference itself can be broken down into the difference between the potential next Q value given some action and the current Q value given the last action taken. The new Q value is the combination of the reward experienced at the current time and the best existing Q value for the next state, determined by iterating through all possible actions at s_{t+1} that can be taken at the current state. This Q value for the next state is discounted by a parameter γ, which is typically less than 1, to lessen the weight of future rewards. The discount value represents the trade-off between obtaining an immediate reward and waiting to get a larger reward in the future. In uncertain environments, the discount value for an agent may be higher, as future rewards are less predictable and guaranteed. In many cases, modelers set γ to 1 or very close to one to make it easier to obtain future rewards.

To show a simple robotic example of Q-learning, suppose that we have a factory robot working on an assembly line. The robot can move left and right along the line, which is split into discrete squares labeled 1 through 5 from left to right. The rightmost square contains a positive reward of 10, such as an item that needs to be picked up, and the leftmost state contains a penalty of −10, such as the presence of another robot that would cause a potential collision. As shown in figure 4.3, the collection of states consists of the five possible locations of the robot and the collection of actions consists of moving one square to the left or right. A table of Q values is initialized to 0, with one row for each possible action and one column for each possible state. The agent starts in the center state and wanders around randomly, applying the Q value update equation at each time step. The Q values do not change until an actual reward is found. At some point during this random walk, the robot will land on a negative reward or a positive reward. For example, if $\alpha=1$ and $\gamma=0.5$, the first time the robot lands on state 4 and makes a right action to state 5, the update equation looks like the following:

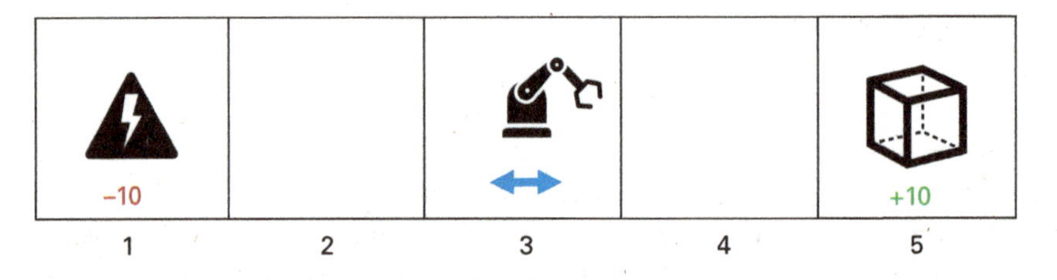

Q table after first transition from state 4 to state 5

	1	2	3	4	5
Left	N/A	0	0	0	0
Right	0	0	0	10	N/A

Q table after first transition from state 5 to state 4

	1	2	3	4	5
Left	N/A	0	0	0	5
Right	0	0	0	10	N/A

Fully trained Q table

	1	2	3	4	5
Left	N/A	−9.33	1.33	2.67	1.33
Right	1.33	2.67	5.33	10.67	N/A

Figure 4.3
A small environment with five states, a negative reward, and a positive reward. Example Q tables after the first transition from state 5 to state 4, after the first transition from state 4 to state 5, and after fully exploring the environment.

$$Q^{new}(4,'right') \leftarrow 0 + 1(10 + .5 * \max_a(Q(4,a) - Q(4,a_t))$$

$$Q^{new}(4,'right') \leftarrow 0 + 1(10 + .5 * 0 - 0)$$

$$Q^{new}(4,'right') \leftarrow 10.$$

The Q table is then updated accordingly. The next time the robot makes a transition going right from state 5 to state 4, the equation is applied as follows:

$$Q^{new}(5,'left') \leftarrow 0 + 1(0 + .5 * \max_a(Q(5,a) - Q(5,a_t))$$

$$Q^{new}(5,'left') \leftarrow 0 + 1\ (0 + .5 * 10 - 0)$$

$$Q^{new}(5,'left') \leftarrow 5.$$

This value is also updated on the table. By applying the learning rule every time step while picking random actions of left and right, the Q values eventually look like the table at the bottom of figure 4.3, with increasing Q values as the agent moves right toward the positive reward and decreasing Q values close to the negative reward. After training, the agent can decide which movements to make on the basis of Q values of each possible action given a state.

Box 4.1 describes Q learning in a Webots simulation in which the robot learns to traverse a double-T maze.

4.4.2 Model-Based Reinforcement Learning

Model-based reinforcement learning requires the extra step of representing state transitions. In the previous example of Q-learning, the state transitions were definite; that is, the action of moving left always caused the state to transition to the state on the left. However, in both model-free and model-based learning, state transitions may contain an element of uncertainty. In this case the state transitions must be represented by probabilities. **Value iteration** is an example of model-based reinforcement learning. Value iteration maintains representations of state transitions, state values, and rewards, all for optimal selection of actions. Much of the terminology and technique used in applying value iteration is similar to applying Q-learning. The equation is described by

$$V^{new}(s_{t+1}) = \max_a\left(\sum_{s'} T(s_t, a, s')[R(s_t, a, s') + \gamma V(s')]\right), \tag{4.2}$$

where the symbol names are the same as in section 4.4.1.

In addition, $V^{new}(s_{t+1})$ is the updated value of state at time $t+1$, s' is the range of possible states, $T(s_t, a, s')$ is the transition probability between state s_t and s' if action a is taken, $R(s_t, a, s')$ is the reward value between state s_t and s' if a is taken.

As in Q-learning, values of each possible state are stored in tables and continually updated, applying the update equation after each action. For example, we could have a robot displaying an emotional state using a facial expression according to its mood. The

Box 4.1

A Webots simulation to examine how model-free reinforcement learning can shape a robot's behavior can be found in the ModelFreeRL_Maze folder on GitHub: https://github.com/jkrichma/NeurorobotExamples/.

In this simulation, the robot navigates a double-T maze (figure 4.4). States 0 through 2 are decision points at the T-junctions where the robot can decide to turn left or right. States 3 through 6 are endpoints where the robot may or may not receive a reward. Q learning is implemented to learn the appropriate actions at each state.

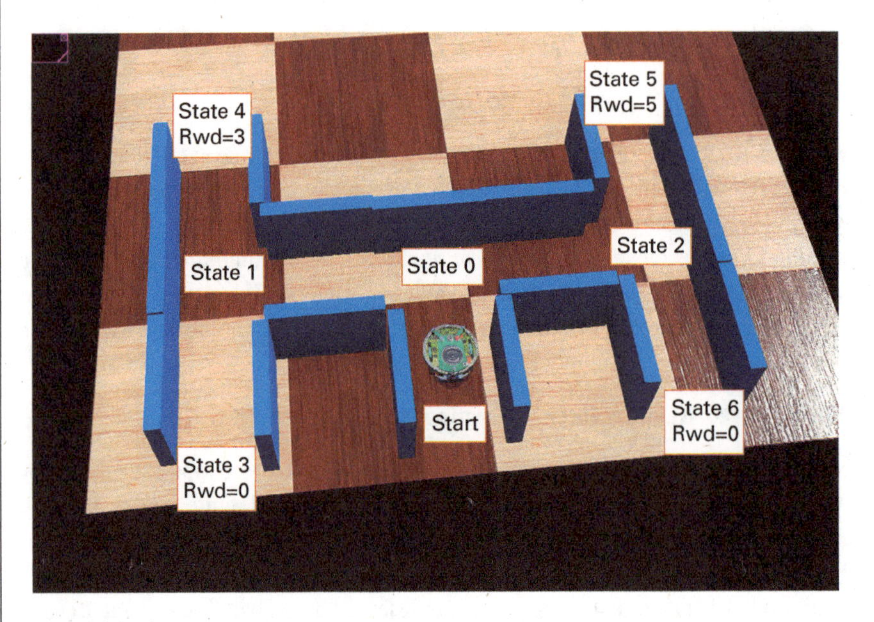

Figure 4.4
Double-T maze using the e-Puck robot. The proximity sensors are used to detect the T-junctions. Supervisor mode in Webots is enabled to know when the robot is at an end state and to return the robot to the start of the maze after each trial.

For example, figure 4.5 shows the robot's ability to get rewards at states 4 and 5. What happens when we move the reward locations for trials 51 to 100? Note that higher beta values in the softmax result (see equation 4.4 in section 4.5.2) in more rewards before the reward locations change, but difficulty adapting after the reward locations change.

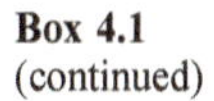

Box 4.1
(continued)

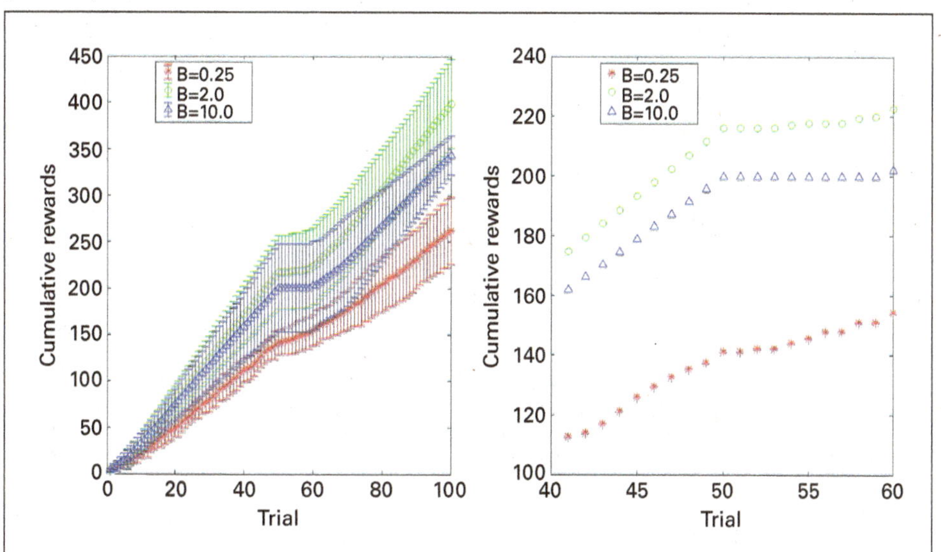

Figure 4.5
The effect of changing the softmax temperature (beta) on getting rewards and responding to reward location changes. The left chart shows the mean and standard deviation over five runs of cumulative rewards received by the robot. The right chart shows the mean cumulative rewards ten trials prior and ten trials after the change for different B values.

Test different values of alpha (learning rate) and beta (softmax temperature). How does changing these parameters affect behavior? What do the state tables and likelihood of turning look like before and after learning? Try different combinations of rewards. How did this change behavior?

three possible moods could be *happy*, *neutral*, and *sad*. The robot could have two possible actions, going toward home and exploring. Every possible pair of state, action, and resulting state would have to be represented, as shown in figure 4.6. These would form the triplets (s_t, a, s_{t+1}). For each triplet, we would store a transition probability in the range of 0 to 1 and a reward value.

4.5 Prediction

In model-based reinforcement learning, the representation of state transition probabilities allows an agent to predict future states, which subsequently allows the agent to plan actions that maximize reward. Prediction is a powerful capability that allows agents to minimize cost and maximize gains over long periods of time. Because the environment is never fully

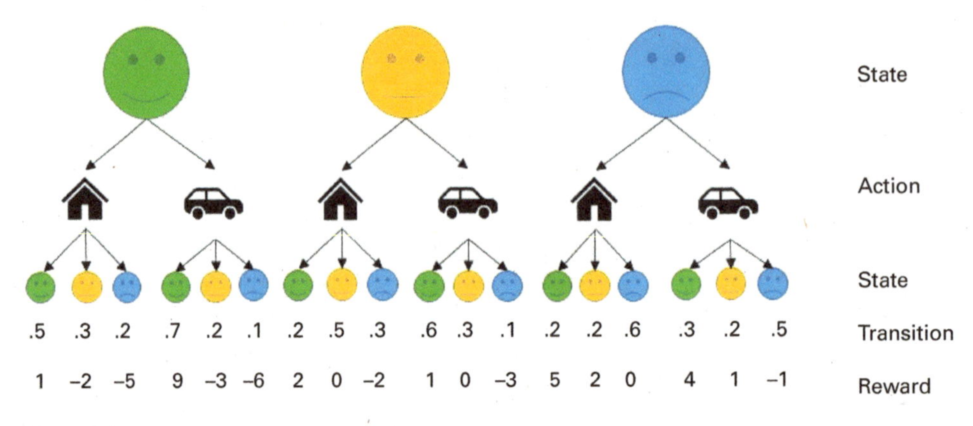

Figure 4.6
A model of state and action transitions. In this example, there are three emotional robot states of *happy*, *neutral*, and *sad*. The action space consists of two actions, *home* and *explore*. When the agent is in one of the three states and selects one of the two actions, the next state is represented as a probability distribution with three possible states. Each possible triplet of state, action, and resulting state has an associated transition probability and reward value.

predictable, uncertainty is often represented as a probability distribution of outcomes. Using these distributions, numerical measurements of predictability can be observed and resulting action selections can depend on the level of predictability.

4.5.1 Measurements of Predictability

Shannon entropy measures the predictability of a probability distribution, and is described by the following equation:

$$H = -\sum_i p_i \log_2(p_i), \tag{4.3}$$

where H is the measurement of Shannon entropy, i is an outcome, p_i is the probability of that outcome, and $\log_2$ is a binary logarithmic function. For instance, suppose there are three coins: a fair coin with equal probability of landing heads or tails, a weighted coin that lands on heads 90 percent of the time, and a weighted coin that lands on heads 10 percent of the time. Changing the value 2 to e, as in the natural log, the three Shannon entropy values of these coins would be calculated

$$H_1 = -(.5(\ln(.5)) + .5(\ln(.5))) = .693$$

$$H_2 = -(.9(\ln(.9)) + .1(\ln(.1)) = .325$$

$$H_3 = -(.1(\ln(.1)) + .9(\ln(.9)) = .325$$

The fair coin has the highest entropy because it is the least predictable. The biased coins have lower entropy, being more predictable. Moreover, their entropy values are the same because the actual outcomes are irrelevant and only the probability distributions are used in the calculation.

In general, a more unpredictable environment requires a greater amount of computation and energy for a robot to operate in that environment. Therefore, even when two actions have the same amount of expected reward, an agent would be more likely to choose the more predictable environment.

4.5.2 Value-Based Action Selection

Uncertainty and variability play a role when selecting actions. The **softmax function** is a means of selecting actions on the basis of a probability distribution of the expected values. If a collection of available actions has a quantity attached to each action, such as a reward value, this can be converted into a probability distribution using the following formula:

$$p_a = \frac{\exp(\beta q_a)}{\sum_{i=1}^{N} \exp(\beta q_i)}, \tag{4.4}$$

where p_a is the probability of taking action a; q_a is the quantity or value associated with action a; N is the number of possible actions; and β is the temperature. The temperature also regulates a trade-off between exploration and exploitation, in which small values of β lead to flatter distributions and more exploration, and larger values of β lead to peakier distributions and exploitation.

Because the softmax function converts the quantities of a distribution to a set of probabilities summing to 1, an action can then be chosen using sampling techniques. Figure 4.7 shows the effects of applying the softmax distribution to a set of four actions. For instance, the four actions could correspond to a robot moving north, south, east, or west, with a reward value associated with each direction, as shown in figure 4.7*A*. Applying the softmax

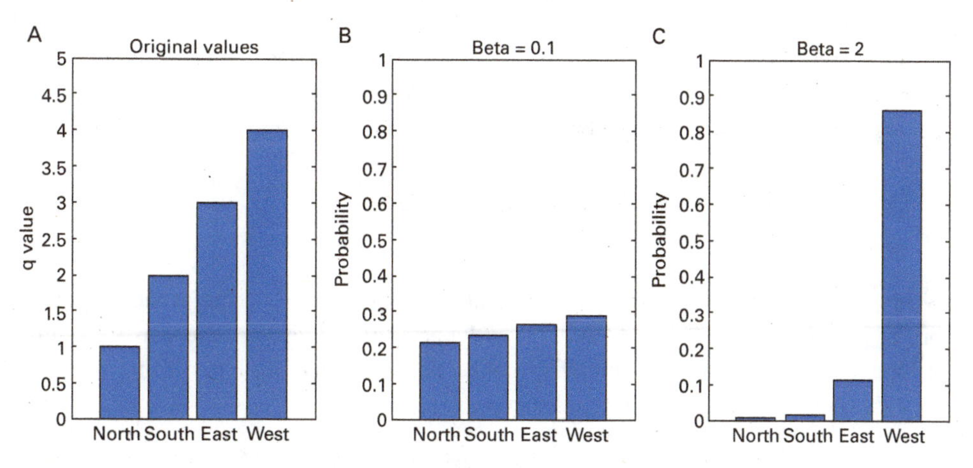

Figure 4.7
Effects of applying the softmax distribution. *Left*: Original quantities for a set of four actions. *Center*: Softmax distribution applied with β equal to 0.1. *Right*: Softmax distribution applied with β equal to 2.

distribution with a small β results in a flat probability distribution with which it is likely for the robot to move in any direction (see figure 4.7*B*). Increasing β causes the probability distribution to be peaky, and there is a high probability that the robot will go west (see figure 4.7*C*). Sometimes β stays constant throughout a simulation. Other times it may be useful to start with a small β to promote exploration and then increase it later in the simulation to exploit learned rewards.

4.6 Case Study: Darwin VII—Perceptual Categorization and Conditioning in a Brain-Based Device

In this case study we describe an early neurorobot, called Darwin VII, with a model that uses unsupervised learning to categorize objects and reinforcement learning to attach value to those objects (Krichmar & Edelman, 2002). Darwin VII autonomously explored its environment and sampled stimuli that contained positive and negative value (see figure 4.8). Through its experience Darwin VII built up perceptual categories of the objects it sampled. Darwin VII's artificial brain was based on the anatomy and physiology of vertebrate nervous systems. The simulated nervous system was made up of a number of areas labeled according to the analogous cortical and subcortical brain regions for vision, auditory processing, and value. Each area contained different types of neuronal units consisting of simulated local populations of neurons or neuronal groups. The simulated nervous system contained eighteen neuronal areas, 19,556 neuronal units, and approximately 450,000 synaptic connections. Figure 4.8*B* shows a high-level diagram of the different neural areas and the synaptic connections between neural areas in the simulated nervous system.

A neuronal unit in Darwin VII was simulated with a mean firing rate model and the activity of such a unit corresponded roughly to the firing activity of a group of neurons averaged over a time period of 200 ms. This corresponded to the time needed to process sensory input, compute neuronal unit activities, update the connection strengths of plastic connections, and generate motor output. The total contribution of synaptic input to unit i is given by

$$A_i(t) = \sum_{j=1}^{N} c_{ij} s_j(t), \tag{4.5}$$

where N is the number of connections to unit i, c_{ij}, is the weight value of the connection projecting to unit i from unit j, and $s_j(t)$ is the activity of unit j at time step t. Negative values for c_{ij} corresponded to inhibitory connections. The activity level of unit i is given by a thresholded hyperbolic tangent function

$$s_i(t+1) = \phi(\tanh(g_i(A_i(t) + \omega s_i(t)))), \tag{4.6}$$

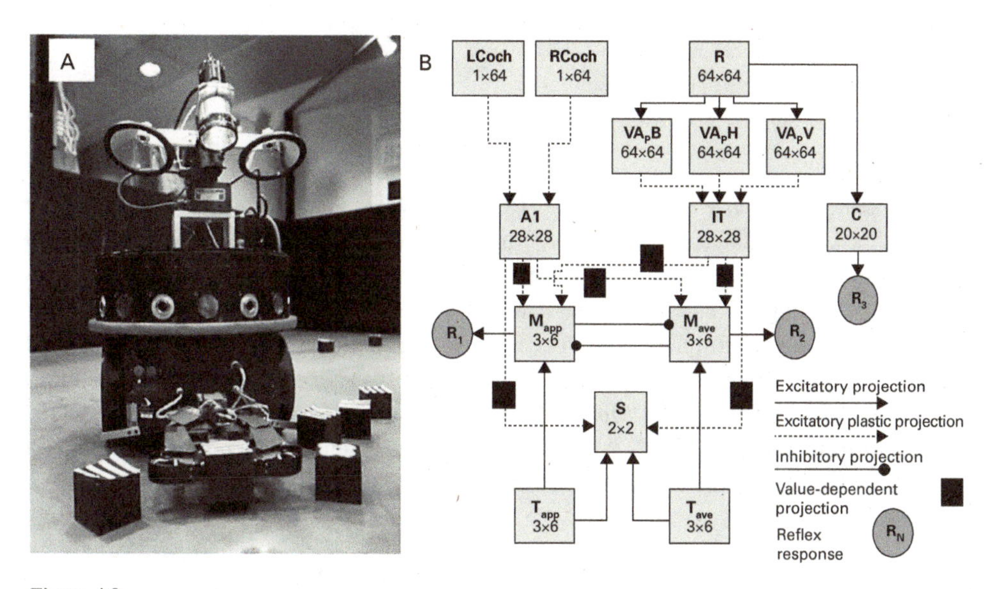

Figure 4.8
A, Darwin VII consists of a mobile base equipped with several sensors and effectors. Darwin VII is constructed on a mobile circular wheeled platform with pan and tilt movement for its camera, and microphones. The CCD camera, two microphones on either side of the camera, and sensors embedded in the gripper, which measures the surface conductivity of stimuli, provide sensory input to the neuronal simulation. *B*, Schematic of Darwin VII's neural architecture. The major systems that make up the simulated nervous system are: an auditory system, a visual system, a taste system, sets of motor neurons capable of triggering behavior, a visual tracking system, and a value system. The 64×64 gray-level pixel image captured by the CCD camera was relayed to a retinal area R and transmitted via topographic connections to a primary visual area VA_P. There were three subpartitions in VA_P selective for blob-like features, for short horizontal line segments, or for short vertical line segments. Responses within VA_P closely followed stimulus onset and projected nontopographically via activity-dependent plastic connections to a secondary visual area, analogous to the inferotemporal cortex (IT). The frequency and amplitude information captured by Darwin VII's microphones was relayed to a simulated cochlear area (L_{Coch} and R_{Coch}) and transmitted via mapped tonotopic and activity-dependent plastic connections to a primary auditory area A1. A1 and IT contained local excitatory and inhibitory interactions producing firing patterns that were characterized by focal regions of excitation surrounded by inhibition. A1 and IT sent plastic projections to the value system S and to the motor areas M_{app} and M_{ave}. These two neuronal areas were capable of triggering two distinct behaviors, appetitive and aversive. The taste system (T_{app} and T_{ave}) consisted of two kinds of sensory units responsive to either the presence or absence of conductivity across the surface of stimulus objects as measured by sensors in Darwin VII's gripper. The taste system sent information to the motor areas (M_{app} and M_{ave}) and the value system (S). Area S projects diffusely with long-lasting value-dependent activity to the auditory, visual, and motor behavior neurons. The visual tracking system controlled navigational movements, in particular the approach to objects identified by brightness contrast with respect to the background. To achieve tracking behavior, the retinal area R projected to area C (colliculus). Adapted from Krichmar and Edelman (2002).

where

$$\phi_i(x) = \begin{cases} 0; & x < \sigma_i \\ x; & \text{otherwise,} \end{cases}$$

and ω is the persistence of unit activity from one cycle to the next, σ_i is a unit specific firing threshold, and g_i is a scale factor, which differed depending on the neural area.

Connections within and between neuronal areas were subject to activity-dependent modification following a value-independent and a value-dependent synaptic rule. Synaptic modification was determined by both presynaptic and postsynaptic activity and resulted in either strengthening or weakening of the synaptic efficacy between two neuronal units. The Bienenstock, Cooper and Munro (BCM) learning rule was used to govern synaptic change because it has a region in which weakly correlated inputs are depressed and strongly correlated inputs are potentiated (Bienenstock et al., 1982).

Value-independent synaptic changes in c_{ij} were given by

$$\Delta c_{ij}(t+1) = \varepsilon(c_{ij}(0) - c_{ij}(t)) + \eta s_j(t) F(s_i(t)), \tag{4.7}$$

where $s_i(t)$ and $s_j(t)$ are activities of postsynaptic and presynaptic units, respectively; η is a fixed learning rate, ε is a decay constant, and $c_{ij}(0)$ is the initial ($t=0$) weight of connection c_{ij}. The decay constant ε governed a passive, uniform decay of synaptic weights to their original starting values. The function F is a piecewise linear approximation of the BCM learning rule.

The synaptic change for value-dependent synaptic plasticity was given by

$$\Delta c_{ij}(t+1) = \varepsilon(c_{ij}(0) - c_{ij}(t)) + \eta s_j(t) F(s_i(t))\overline{S}, \tag{4.8}$$

where $\overline{S}$ is the average activity of the value system S (see figure 4.8*B*).

Darwin VII's environment consisted of an enclosed area with black walls and a floor covered with opaque black plastic panels, on which metallic cubes were distributed (see figure 4.8*A*). The top surfaces of the blocks were covered with black and white patterns: blobs and stripes. Stripes on blocks in the gripper can be viewed in either horizontal or vertical orientations, yielding a total of three stimulus classes of visual patterns to be discriminated (blob, horizontal, and vertical). A flashlight mounted on Darwin VII and aligned with its gripper caused the blocks, which contained a photodetector, to emit a beeping tone when Darwin VII was in the vicinity. The sides of the stimulus blocks were metallic and could be rendered either strongly conductive ("good taste" or appetitive) or weakly conductive ("bad taste" or aversive). Gripping of stimulus blocks activated the appropriate taste neuronal units (either area T_{app} or area T_{ave}) to a level sufficient to drive the motor areas above a behavioral threshold. In the experiments, strongly conductive blocks with a striped pattern and a 3.9 kHz tone were chosen arbitrarily to be positive-value exemplars, whereas weakly conductive blocks with a blob pattern and a 3.3 kHz tone represented negative-value exemplars.

Early during the conditioning trials, Darwin VII picked up and "tasted" blocks that led to either appetitive or aversive responses (see figure 4.9*A*). During this period, it was the output of the taste neuronal units that activated the value system (S) and drove the motor neuronal units (M_{app} and M_{ave}) to cause a behavioral response. After conditioning, however, both the value system and the motor neuronal units were immediately activated upon the onset of IT's response to a visual pattern or A1's response to a tone. This shift from value system activity that was triggered in early trials by the unconditioned stimulus to value system activity triggered at the onset of the conditioned stimulus is analogous to the shift in dopaminergic neuronal activity found in the primate ventral tegmental area after conditioning (Schultz et al., 1997).

After associating visual patterns with taste, Darwin VII continued to pick up and "taste" stripe-patterned blocks, but it avoided blob-patterned blocks (see figure 4.9*B*). After associating auditory sounds with taste, Darwin VII continued to pick up the high-frequency beeping blocks, but avoided the low-frequency beeping blocks (see figure 4.9*C*).

In Darwin VII, activity in the simulated inferotemporal cortex, IT, provided the basis for visual perceptual categorization. Initially IT's responses to visual stimuli were weak and diffuse (see IT activity shown in figure 4.9*A*). After approximately five stimulus encounters, activity-dependent plasticity between the primary visual cortex, VA_P, and IT caused IT responses to the different stimuli to become strong, sharp, and separable (see IT activity shown in figure 4.9*B*).

Darwin VII's object recognition was observed to be invariant with respect to scale, position, and rotation (see figure 4.10). Note how the object in Darwin VII's field of view changes position and orientation, as apparent in the response of VA_PH (top row of figure 4.10). The neuronal units in VA_PH are retinotopically mapped to the pixels from Darwin VII's camera. In contrast, area IT was a nontopographic self-organizing map. The persistence parameter, ω, of IT neuronal units was higher than those in area VA_PH, which led to a stable activity pattern in IT as an object moved across Darwin VII's field of view. The competition among activity-dependent plastic connections between VA_P and IT as described previously led to different activity patterns for each object. Because of these neural dynamics, plasticity, and seeing the world as a stream of information, visual categorization of a stimulus occurred no matter where an object appeared in Darwin VII's visual field. The apparent size of the stimulus ranged from a maximum when the object was directly in front of Darwin VII to one-quarter of the maximum size when the object was far from Darwin VII. Correct categorization of striped blocks in Darwin VII's field of vision occurred when blocks were horizontal or vertical in its gripper or when the stripes on the blocks were rotated over a range of ±30° of a horizontal or vertical reference as Darwin VII approached the blocks.

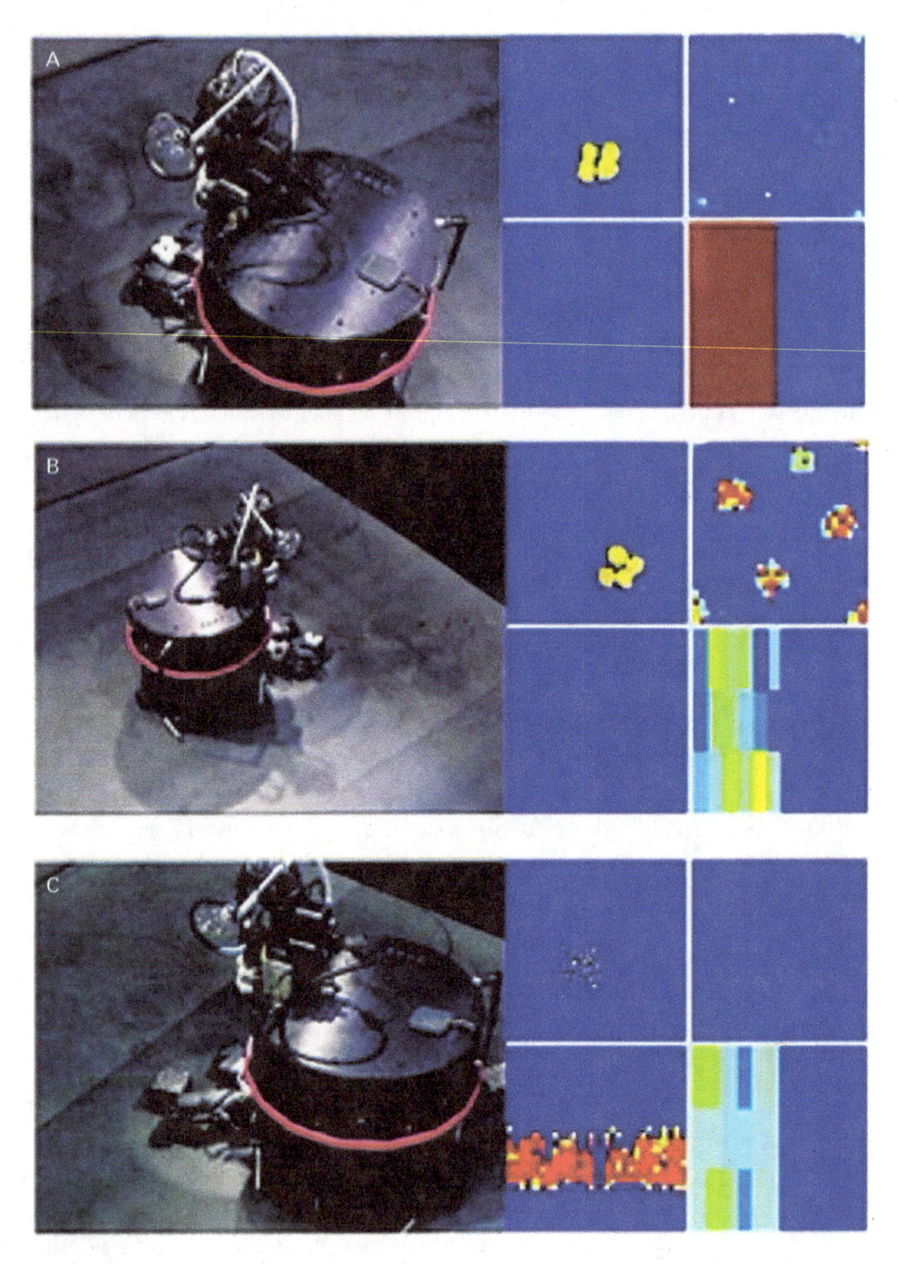

Figure 4.9
Darwin VII during behavioral experiments. The panels to the right of Darwin VII show activity of selected neural areas in the simulation (R, *top left*; IT, *top right*; A1, *bottom left*; M_{ave}, *bottom right, left side*; M_{app}, *bottom right, right side*). Each pixel in a selected neural area represents a neuronal unit, and activity is normalized in a range from no activity (dark blue) to maximal activity (bright red). *A*, Darwin VII upon the first encounter of an aversive block. In this early conditioning trial, Darwin VII is shown picking up and "tasting" an aversive block. Activity in IT is insufficient, but activity in the taste system T_{ave} is sufficient to drive activity in the aversive motor behavior neural area (M_{ave}) above the behavioral threshold. *B*, Darwin VII upon the tenth encounter with an aversive block having blob visual patterns. After primary conditioning with visual stimuli, activity in area IT is sufficient to drive the M_{ave} neuronal units above the behavioral threshold, triggering a motor response to avoid "tasting" an aversive block. *C*, Darwin VII upon the tenth encounter with an aversive block having only auditory cues. After primary conditioning with auditory stimuli, activity in area A1 is sufficient to drive the M_{ave} neuronal units above the threshold to trigger a behavioral response.

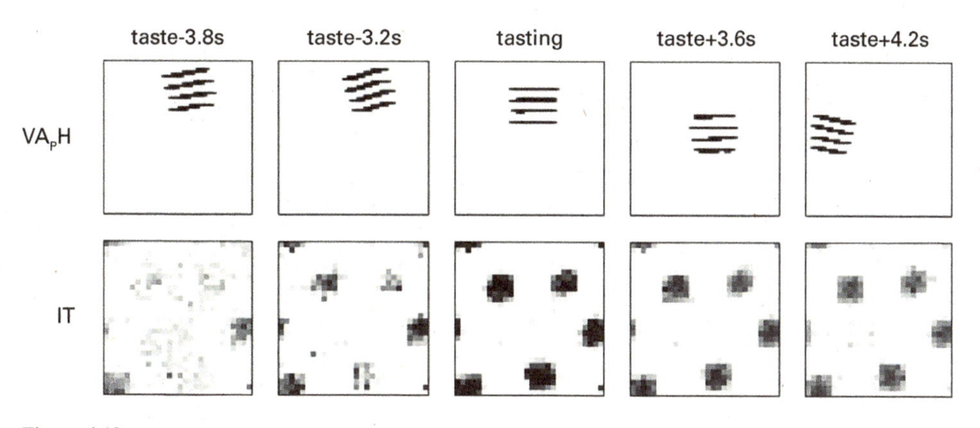

Figure 4.10
Invariant object recognition in Darwin VII. *Top row*: Primary visual area for horizontally oriented lines (VA_pH). VA_pH has a retinotopic map, in which each pixel represents the receptive field of a neuronal unit. *Bottom row*: Simulated inferotemporal cortex (IT). IT is a self-organizing map with no retinotopy. Each grayscale pixel shows the activity in area ranging from no activity (white) to maximal activity (black).

Darwin VII is a conditioning experiment in which the taste of a block is the unconditioned stimulus (US), and the approach behavior and avoid behavior are the unconditioned responses (UR). The sound and the visual object categories are the conditioned stimuli (CS) and the conditioned responses (CR) are the same as the UR, but driven by the CS (auditory or visual stimuli). Figure 4.11 shows the percentage of CRs driven by the auditory or visual stimulus, for seven Darwin VII trials. The value-based learning in Darwin VII can be thought of as a neural implementation of model-free reinforcement learning. The low-frequency tone, high-frequency tone, blob pattern, horizontal pattern, and vertical pattern are states. The appetitive approach to an object and the aversive avoidance of an object are actions. The salience area S provides a value signal that leads to a state-action mapping. The increase in conditioned response showed that Darwin VII learned that auditory or visual cues predicted the value of the object, which resulted in its taking the appropriate behavioral response. These learning curves closely resembled those for similar conditioning experiments in rodents, pigeons, and other organisms.

The behavior of Darwin VII showed that a robot operating on biological principles and without prespecified instructions could carry out perceptual categorization and conditioned responses. In both the perceptual categorization and the conditioning experiments, the development of categorical responses required exploration of the environment and sensorimotor adaptation through specific and highly individual changes in connection strengths. Darwin VII laid groundwork for increasingly sophisticated neurorobots with more complex neural circuits and morphologies, which gave further insights into the relationships between brain, body, and behavior.

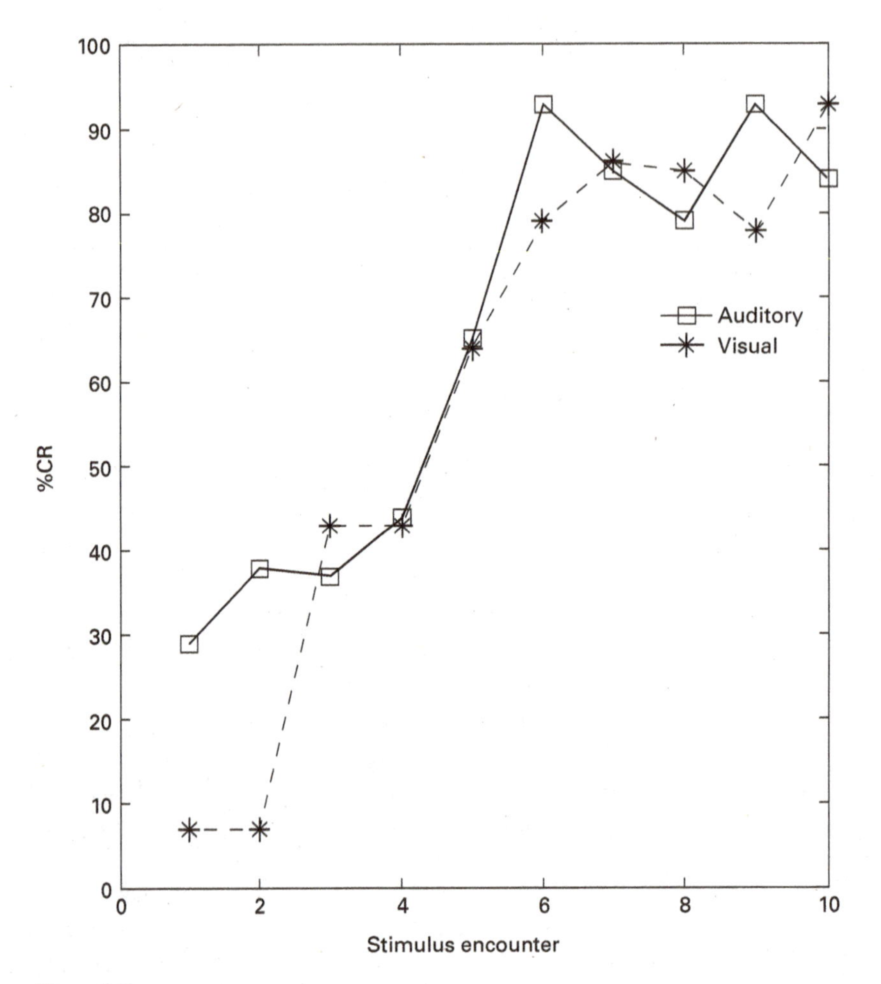

Figure 4.11
The percentage of conditioned responses (%CR) per stimuli encountered by Darwin VII for auditory and visual stimuli. Each point is the average %CR for seven Darwin VII trials.

4.7 Summary and Conclusions

In this chapter we covered techniques for robots operating in unpredictable environments. Representing the problem as a Markov decision process (MDP) formalizes the states, actions, and rewards of the system, allowing robotic agents to calculate actions that maximize their rewards using reinforcement learning. Model-free reinforcement learning determines ideal actions for each state without explicitly representing state transition probabilities, whereas model-based reinforcement learns state transitions, predicting future states based on actions taken in present states. The predictability of states can be measured using Shannon entropy, and actions can be sampled from distributions using the softmax function.

The Darwin VII case study brings together the following five concepts that were presented in previous chapters. (1) The neural network architecture, consisting of the different brain areas and connections between areas, was strongly influenced by mammalian neuroanatomy. (2) The neuronal units were implemented with a mean firing rate model using a hyperbolic tangent activation function. (3) The categorization of objects in Darwin VII's world were learned with an unsupervised BCM learning rule. (4) The experimental design was based on Pavlovian conditioning. Note that because learning was expressed by Darwin VII's actions, the paradigm is called operant conditioning. (5) Reinforcement learning led to Darwin VII approaching good-tasting blocks and avoiding bad-tasting blocks.

II NEUROROBOT DESIGN PRINCIPLES

In their book *How the Body Shapes the Way We Think: A New View of Intelligence*, Pfeifer and Bongard (2006) put forth an embodied approach to cognition. Because of this position many of the robots that they designed demonstrate "intelligent" behavior with limited neural processing (Bongard et al., 2006; Iida and Pfeifer, 2004). It is our belief that neurorobots should attempt to follow many of these principles. In this part we discuss a number of principles to consider in designing neurorobots and using them to test brain theories. The chapters in this part are strongly inspired by Pfeifer and Bongard's design principles for intelligent agents. We build on these design principles by grounding them in neuroscience and by adding principles based on neuroscience research.

5 Neurorobot Design Principles 1: Every Action Has a Reaction

5.1 Introduction

In this first set of neurorobotic design principles, we focus on what Pfeifer and Bongard (2006) called the "here and now." The design principles discussed in this chapter are grounded in neuroscience and are focused on processes that respond to events. Even without learning and memory, these processes lead to flexible, adaptive behavior. We begin with the recurring theme for this book on neurorobotics, which is that the brain needs a body (Chiel & Beer, 1997).

5.2 Embodiment

Brains do not work in isolation; they are closely coupled with the body acting in its environment. *The brain is embodied, and the body is embedded in the environment.* Biological organisms perform **morphological computation**; that is, certain processes are performed by the body that would otherwise be performed by the brain (Pfeifer and Bongard, 2006). Moment-to-moment action can be handled at the periphery by the body, sensors, actuators, and possibly reflexes at the spinal cord level. This allows the central nervous system, which is slower and requires more processing than the body or peripheral nervous system, to predict, plan, and adapt by comparing its internal models with current information from the body (Hickok et al., 2011; Shadmehr & Krakauer, 2008).

In an interesting paper, "What's that thing called embodiment?," Tom Ziemke (2003) discussed the following six different notions of embodied cognition. (1) Structural coupling between agent and environment. This is related to the three-constituent design principle from Pfeifer and Bongard. The design of an embodied agent must consider its behavior, body morphology, and ecological niche, that is, the specific environmental conditions that fit the agent's needs. For example, human vision samples wavelengths in the environment that are different from the wavelengths sampled by a honeybee. (2) Historical embodiment due to adaptation through agent-environment interaction. As discussed in the next chapter, organisms can adapt, learn, and remember. What is learned and how the organism adapts are closely tied to the organism's experience in the environment. (3) Physical embodiment with sensors and actuators. The range, resolution, and layout of an organism's sensors as well as the range and layout of the organism's actuators strongly shape behavior. (4) Organism-like bodily form also shapes behavior. (5) Embodiments of living systems act on their own and are not programmed. Finally, (6) Social embodiment. We consider social embodiment in a later chapter.

Let's consider embodiment notions 3, 4, and 5 together with an illustrative example. Think of a toddler flailing his or her arms. The child's arms more easily move toward the front of the torso than the back. By chance the toddler's hand touches an object causing a reflexive grasping motion. This leads to the fingers, where most of the tactile sensors are located, touching the object. The child will then move this hand in the easiest direction, which tends to be toward the face, where a range of vision, olfaction, and taste receptors reside. The embodiment of this "autonomous system" was facilitated by the shape of the toddler's body and the layout of its sensors and actuators. These notions can provide guidance for the design of neurorobots. The organism's body form shapes its behavior and its brain function. This requires a brain and body that is engaged with the environment.

5.2.1 Close Coupling of Brain and Body with Ecological Niche

A classic example of how an organism's brain and body is closely tied to its ecological niche is cricket phonotaxis. A male cricket generates a mating song that has a particular pitch and frequency. The female cricket can distinguish and locate the source of this sound. Barbara Webb's group designed a neurorobot to show how crickets might achieve this task (Webb & Scutt, 2000). In this system, a combination of a biologically plausible spiking neural network and the appropriate layout of auditory sensors led to a simple yet elegant solution to phonotaxis (see figure 5.1). Using only four neurons in their neural network, they were able to reproduce the behavior of the cricket approaching a mate's song. The directional response depended on relative latencies in firing onset of the neurons rather than the firing rate. The layout of the ears, the properties of the ears, and the properties of the neurons were all highly tuned to the environmental signature of the cricket song. In the neurorobot and in the real cricket, this close coupling of the brain, body, and environmental niche for the cricket was key for successful behavior.

5.2.2 Trapping a Soccer Ball: Body and Peripheral Sensing Frees Central Processing

For many tasks that we carry out with ease, our brains are too slow to sense, process and move. For example, skiing down a hill or catching a wave with a surfboard happens too fast for our central nervous system to position the body appropriately. But the form and compliance of the body can automatically position itself properly without brain control and adjust itself to perturbations. This is morphological computation in action.

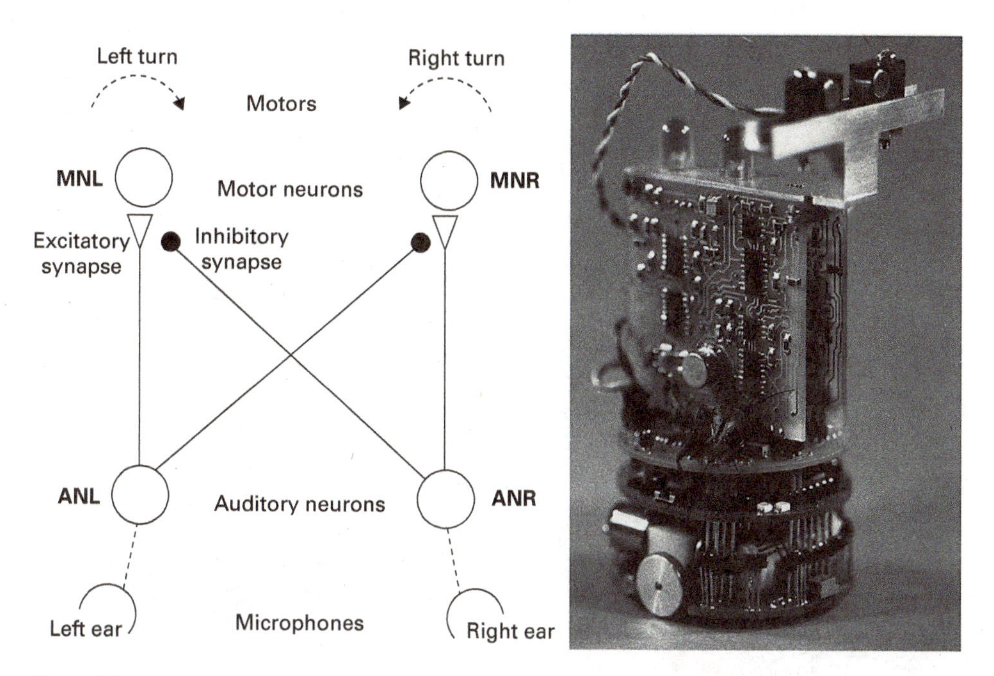

Figure 5.1
Neurorobotic model of cricket phonotaxis. *Left*: Auditory neurons (AN), modeling the AN pair in the cricket prothoracic ganglion, receiving auditory input. Each connects to a motor neuron (MN). By exciting the MN on the opposite side of the sound and inhibiting the MN on the side of the sound, the robot will move toward the sound source. *Right*: Khepera robot with microphones to mimic the cricket. Adapted from Webb and Scutt (2000).

Trapping a soccer ball is another example of morphological computation. In a RoboCup tournament, the Segway soccer team from The Neurosciences Institute solved this difficult sensorimotor problem with a very cheap design (Fleischer et al., 2006). On a large playing field it was nearly impossible for the robot to catch a fast-moving soccer ball, given that the Segway was large and cumbersome and had a slow camera frame rate. Soccer balls would bounce off the robot before it had a chance to respond. After much trial and error, the team used plastic tubing that was fastened around the robot's body like a hula-hoop at just the right height (see figure 5.2*E*). Any ball that was passed to the Segway robot was trapped by the tubing, giving the robot time to use its camera and proximity sensors to place the ball in its kicking apparatus. In a sense, this is what humans do when playing soccer. They use soft pliant materials angled appropriately to soften the impact of a ball coming toward them. Many actions like these take place without much thought (i.e., brain processing).

There is an important lesson in neuroscience that can be learned from these embodied neurorobot examples. By putting more emphasis on designs that exploit the environment, we can offload some of the control from the cognitive robot's central nervous system onto the body itself. This should allow the robot to be more responsive to the environment and

Figure 5.2
Segway soccer robots. *Left*: Segway scooter adapted for soccer. *Right*: Soccer-playing robot built on a Segway base. *A*, Active capture devices. *B*, Laser rangefinder. *C*, Pan-tilt unit and camera. *D*, Kicking assembly. *E*, Passive capture ring. *F*, Voice command module. *G*, Crash bars. From Fleischer et al. (2006).

be more fluid in its actions. In addition, it frees up the nervous system to put more emphasis on planning, prediction, and error correction rather than reflexive movements. Too often cognitive neuroscientists forget that the body and the peripheral nervous system are performing many vital, moment-to-moment behaviors and tasks without central control.

5.3 Efficiency through Cheap Design

Constructing agents, which are built to exploit properties of the ecological niche, can make many tasks easier or *cheaper*. Often this design principle is mistakenly thought of as using the least expensive materials to construct the robot. This is certainly a worthy goal, but what is really meant by *cheap* is the efficient use of resources. To paraphrase the Occam's razor principle, cheap design means finding the simplest solution to the challenge the robot is facing. One way to do this is by exploiting the environment. For example, winged insects and fast swimming fish exploit their environment by creating vortices with their wing beats or fin movements, which causes additional thrust and more energy to come out than the animal put in. Most of our touch receptors are where we need them most, at our fingertips. It would be wasteful to have this fine level of resolution on the back of our hands or arms. This is what we mean by *cheap*.

5.3.1 The Body's Cheap Design

Take bipedal walking, for example, or legged locomotion in general. For years, roboticists have tried and failed to create robust controllers for legged locomotion. The importance of the cheap design principle can be observed when comparing the biped locomotion of passive dynamic walking robots to sophisticated humanoid robots, such as Honda's Asimo or Aldebaran's NAO. Passive dynamic walking robots exploit gravity, friction, and the forces generated by their swinging limbs (Bhounsule et al., 2014; Collins et al., 2005; Collins & Ruina, 2005). As a result, they require very little energy or control to move (see figure 5.3). In contrast, robots such as Asimo need complex control systems and long-lasting batteries to achieve the same result. Although these passive walkers are not necessarily biologically inspired, once the engineers or artists implement a design that minimizes energy expenditure, the gait looks very natural (Andy Ruina, personal communication). Saving energy is a recurring theme in biology (Beyeler et al., 2019; Krichmar et al., 2019). It is also a trade-off. For example, it is more efficient to walk on four legs, but then arms are not available for manual dexterity and gestures, which is important for bipedal organisms.

5.3.2 The Brain's Cheap Design

It is not only the body that follows the principle of cheap design: brains do as well. Biological systems are under extreme metabolic constraints and need to represent information efficiently. Therefore the nervous system must encode information as cheaply as possible. Many of the design principles mentioned previously can be recast in light of this statement. The brain operates on a mere 20 watts of power, approximately the same power required for a ceiling fan operating at low speed (Krichmar et al., 2019). Although being severely metabolically constrained is at one level a disadvantage, evolution has optimized brains in ways that lead to incredibly efficient representations of important environmental features that are distinctly different from those employed in current digital computers. The human brain has many means to reduce its functional metabolic energy use. Indeed, one can observe at every level of the nervous system strategies to maintain high performance and information transfer while minimizing energy expenditure.

At the neuronal coding level, the brain uses several strategies to reduce neural activity without sacrificing performance. Neural activity (i.e., the generation of an action potential, the return to resting state, and synaptic processing) is energetically very costly, and this can

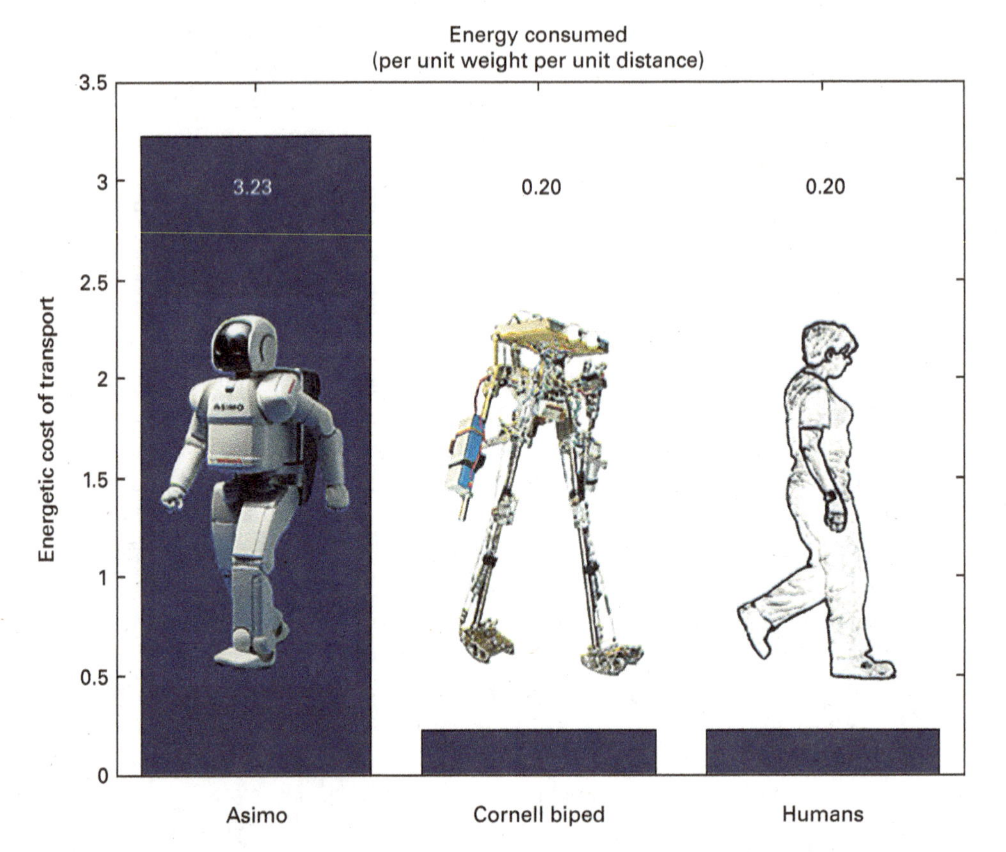

Figure 5.3
Cost of transport comparison. The normalized energy consumed by the Cornell biped robot was similar to humans and less than one tenth that of Honda's Asimo. It used less energy per unit distance per unit weight than any other walking robot to date. Image by Steve Collins (https://www.andrew.cmu.edu/user/shc17/Robot/Robot_photos.htm).

drive the minimization of the number of spikes necessary to encode the neural representation of a new stimulus. Such sparse coding strategies appear to be ubiquitous throughout the brain (Beyeler et al., 2019). Efficient coding reduces redundancies and rapidly adapts to changes in the environment (see figure 5.4). At a macroscopic scale, the brain saves energy by minimizing the wiring between neurons and brain regions (i.e., number of axons) and yet still communicates information at a high level of performance (Laughlin & Sejnowski, 2003). Information transfer between neurons and brain areas is preserved by a small-world network architecture, which reduces signal propagation (Sporns, 2010). These energy-saving ideas should be taken into consideration in constructing neural controllers for robots, which like biological organisms have limited energy resources. Many of these strategies could inspire new methodologies for constructing power-efficient artificial intelligence.

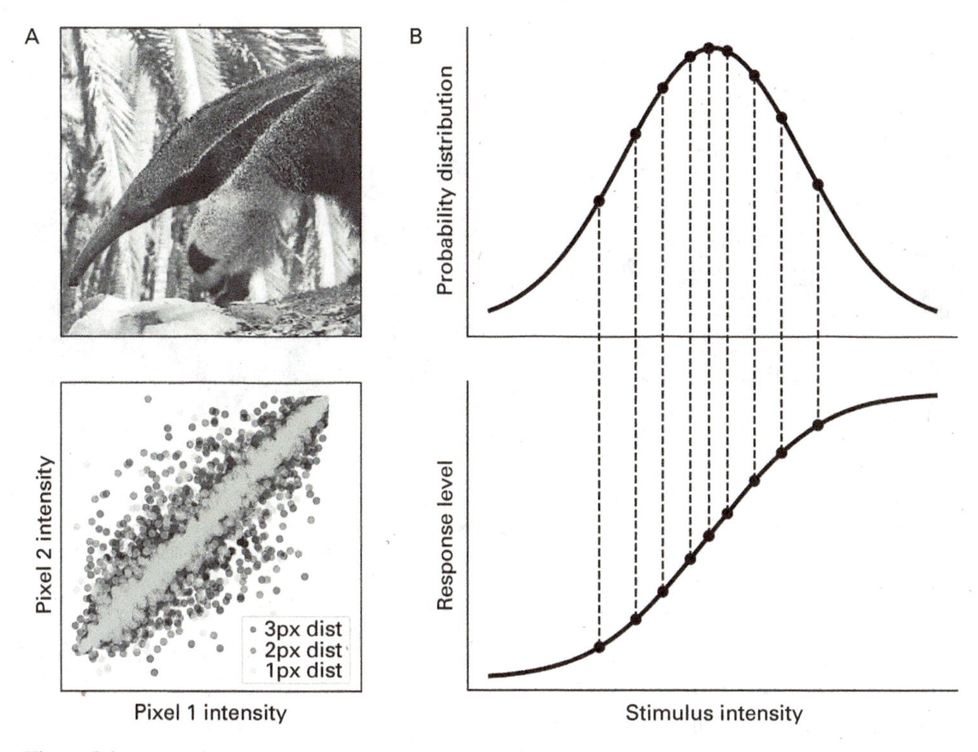

Figure 5.4
Efficient coding. *A*, Sensory stimuli in the environment, such as an image of an anteater, display significant statistical structure. For example, the luminance value of nearby pixels in the image is significantly correlated, an effect that exists even for nonadjacent pixels. Neural systems can improve their coding efficiency by reducing such redundancy. *B*, For a given distribution of sensory characteristics in the world (*top*), a neuron's information capacity is maximized when all response levels are used with equal frequency (*bottom*). Intervals between each response level encompass an equal area under the intensity distribution, so each state is used with equal frequency.

5.4 Sensory-Motor Integration

In the brain, the sensory and motor systems are tightly coupled. An organism or robot may get new sensory information that causes an action. Each action then creates new sensory information. Neurorobots can take advantage of this tightly coupled loop. For example, **figure-ground segregation** is a difficult computer vision problem in which a scene of static objects needs to be recognized from the background. However, segmentation can be facilitated if a hand happens to push an object, resulting in the object moving with respect to its background (Fitzpatrick & Metta, 2003). In this way the motor system can generate sensory information (see figure 5.5). In the nervous system a copy of the action, called a **motor efference copy**, is fed back to the brain. It creates an expectation that can be used to error check the movement or, as shown in figure 5.5, an expectation of a sensory experience.

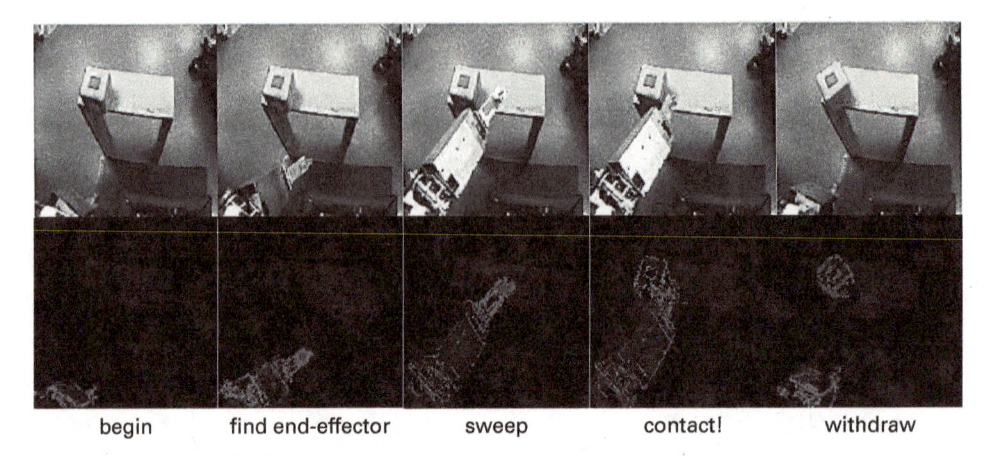

Figure 5.5
The upper sequence shows an arm extending into a workspace, tapping an object, and retracting. This is an exploratory mechanism for finding the boundaries of objects; it essentially requires the arm to collide with objects under normal operation rather than as an occasional accident. The lower sequence shows the shape identified from the tap using simple image differencing. From Fitzpatrick & Metta (2003).

Because hitting the object created a violation of both tactile and visual sensory expectations, the object was easily differentiated from the table.

It is important to emphasize how much the sensory and motor nervous systems are intertwined. Too often neuroscientists study these systems separately, but they are highly interconnected and work in concert. Although there are brain areas specialized for sensing such as auditory cortex and visual cortex, and there are areas of the brain devoted to action such as the motor cortex, most of the cortex is associational and cannot be called exclusively sensory or motor systems. Figure 5.6 shows how interconnected these areas are and how the delineation between perception and action becomes blurred as you move away from the primary sensory and primary motor cortex (Fuster, 2004). The parietal cortex receives **multimodal** sensory inputs and is important for planning movements. By multimodal we mean that the brain area receives more than one sense: auditory, visual, touch, or vestibular. The frontal cortex also receives multimodal inputs and is important for decisions, control of actions, and action selection. These multimodal association areas have direct influence on what we perceive and how we move.

5.5 Degeneracy

Degeneracy is the ability of elements that are structurally different to perform the same function or yield the same output (Edelman & Gally, 2001). To be fault tolerant and flexible a system's architecture should be designed such that different subsystems have different functional processes and there is an overlap of functionality between subsystems.

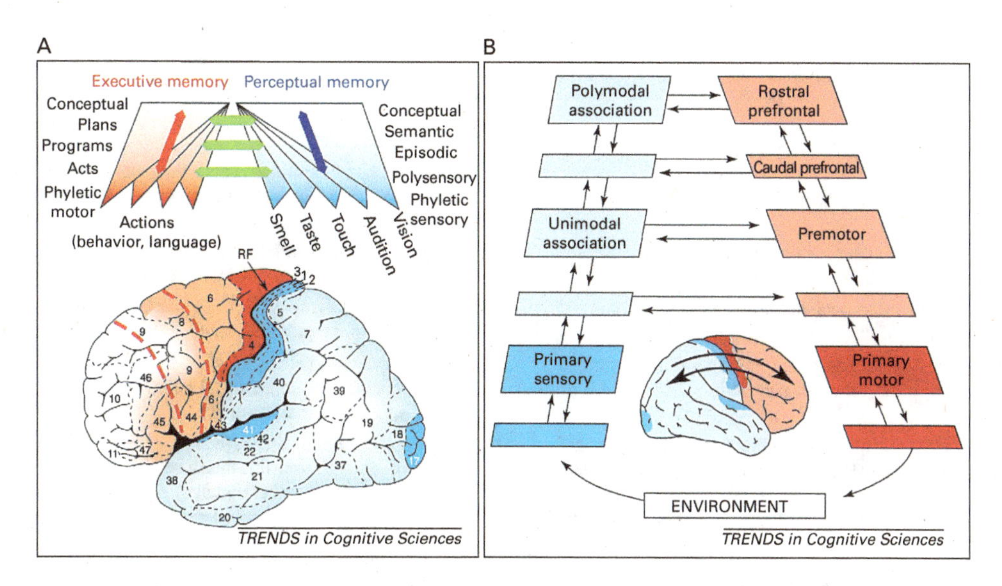

Figure 5.6
Representational map of the human lateral cortex. *A*, Schema of the hierarchical organization of memory and knowledge. *B*, Approximate topographic distribution of memory networks, using the same color code as in *A*. Blue represents the perception side of the cycle, and red the action side. Unlabeled rectangles represent intermediate areas or subareas of labeled cortex. Adapted from Fuster (2004).

In this design, if any subsystem fails the overall system can still function. This is different from redundancy, in which an identical system copy is kept in case there is a system failure (e.g., redundant array of independent disk [RAID] computer memory systems). Degeneracy shows up throughout biology, from low-level processes such as the genetic code and protein folding to system-level processes such as behavioral repertoires and language (see table 5.1). For example, there are four nucleotide bases in DNA (thymine, cytosine, adenine, and guanine). It takes three bases to encode an amino acid, which is the building block of proteins. This means that there are potentially 4^3 or sixty-four possible combinations, but only twenty amino acids make up the proteins found in the human body. In many cases different triplets encode the same amino acid. Therefore the genetic code is considered degenerate. As a result, the genetic code is fault tolerant to mutations. At the other end of the biological spectrum is communication. We have an almost infinite number of ways to communicate the same message. The same message could be communicated through voice, text, email, Morse code, gesture, or facial expressions.

Degeneracy at multiple levels was nicely demonstrated by the neurorobot Darwin, which was used to demonstrate spatial and episodic memory (Fleischer et al., 2007; Krichmar, Nitz, et al., 2005). Darwin solved a dry version of the **Morris water maze** and a place-learning version of a plus maze (see figure 5.7). In the Morris water maze, a rat swims through murky water until it finds a platform hidden beneath the surface. After several days

Table 5.1
Degeneracy at different levels of biological organization

Cellular level	Organism level
Genetic code (many different nucleotide sequences encode a polypeptide)	Connectivity in neural networks (the local circuitry, long-range connections, and neural dynamics varies within and between organisms)
Protein folding (different polypeptides can fold to be structurally and functionally equivalent)	Sensory modalities (information obtained by any one modality often overlaps that obtained by others)
Food sources and end products (an enormous variety of diets are nutritionally equivalent)	Body movements (many different patterns of muscle contraction yield equivalent outcomes)
Cells within tissues (no individual differentiated cell is uniquely indispensable)	Behavioral repertoires (many steps in stereotypic feeding, mating, and other social behaviors are either dispensable or substitutable)
Immune responses (populations of antibodies and other antigen-recognition molecules are degenerate)	Interanimal communication (there is a nearly infinite number of ways to transmit the same message, a situation most obvious in language)

of exploration, the rat will swim directly to the platform from any starting location. If the hippocampus is damaged, the rat cannot learn the location of the platform. In the dry version, a reflective piece of paper was the proxy for the platform. It was the same color as the floor, so the robot could not see the platform but it could feel when it was on the platform by means of a downward-pointing light sensor.

Darwin had an extensive model of the hippocampus formation and its surrounding cortical regions. Similar to a rodent, as the robot explored its environment, hippocampal place cells emerged. Place cells are like the brain's GPS system. A place cell will be active in a specific location of the environment. In the rat and the Darwin experiments, combinations of place cells were used to plan routes to goals. The robot's behavior and neural activity was directly compared with rodent experiments. Like the rat, place cells could not only be used to predict the current location of the robot but also the location from which the robot came and the location to which it was heading. In the robot experiments, several levels of degeneracy emerged.

5.5.1 Degeneracy at the Neuronal Level

Because the neuroroboticists were able to track every neuron in its simulated nervous system, they were able to trace the neuronal activity that led to hippocampal place activity. Although hippocampal place activity was similar on different trials when the robot passed through the same location on the same heading, the neuronal activity leading to that neuron's place activity on a given trial differed dramatically. That is, different neural activation patterns led to the same hippocampal place cell outcome.

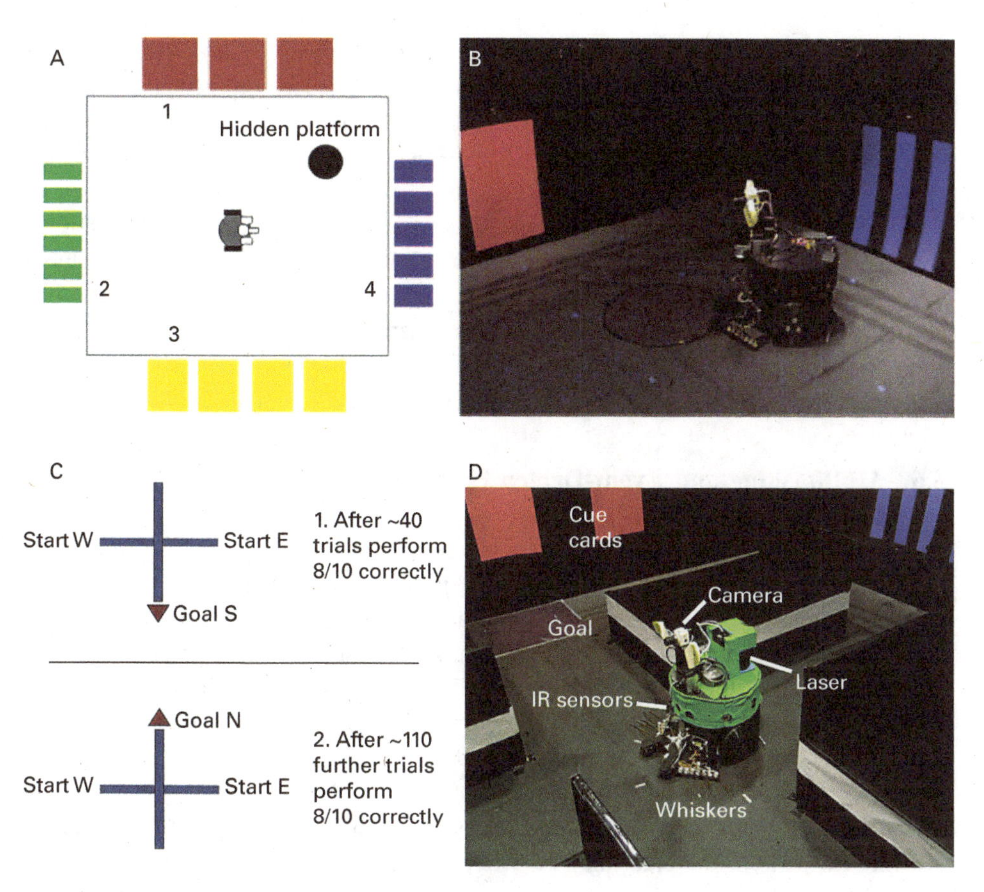

Figure 5.7
Darwin robot experiments for spatial and episodic memory. *A*, Experimental setup for a dry variant of the Morris water maze spatial memory task. *B*, Physical instantiation of the spatial memory task. *C*, Experimental setup for a standard plus maze for place learning. *D*, Physical instantiation of the plus maze.

5.5.2 Degeneracy at the System Level

Darwin received sensory input from its camera (vision), whiskers (somatosensory), compass (head direction), and laser range finder (depth/distance). Darwin's spatial memory was multimodal and degenerate. Even when one or more of its sensory modalities were lesioned, Darwin's behavior and place cell activity remained stable. Different sensory pathways led to the same outcome of knowing where Darwin was located.

5.5.3 Degeneracy at the Individual Level

In the Morris maze task, nine different Darwin *subjects*, which consisted of the identical robots with slightly different nervous systems due to variations in initial synaptic connection probabilities, solved the same spatial navigation task in unique ways. Some subjects bounced off the "red" wall to the hidden platform, some bounced off the "blue" wall, and others went directly toward the platform location. The proficiency of each subject differed as well. Some were better learners than others. However, despite their idiosyncrasies they all shared the same outcome of solving these tasks.

Experience in the real world has a strong shaping effect on brain and behavior. For this reason one should always run multiple subjects on a behavioral task, even if the subjects are neurorobots. In every experiment we have run no two neurorobots have been alike, even when the device and nervous system were identical. In the case of Darwin VII, it was shown that even when the initial weights and connections were the same, experience shaped the way the robot perceived the world (Krichmar & Edelman, 2002).

5.6 Multitasking and Event-Driven Processing

Cognitive scientists tend to study the brain in a serialized fashion by focusing on one subsystem at a time, be it a type of memory or a specific perceptual effect. But intelligence emerges from many processes operating in parallel and driven by events. We (humans and other organisms) are multitaskers, and to multitask we must do things in parallel. The brain is the ultimate event-driven, parallel computer. There is no overarching clock as in computer architectures. Rather the brain responds to events when they happen. Therefore, neurorobots should have a multitask design that responds to multiple, asynchronous events in a timely manner.

Neurons throughout the brain are responding simultaneously to multiple events. Although the different parts of the brain may be acting somewhat independently, they are highly interactive (see figure 5.6). The sensory system is telling the motor system what it senses, and the motor system is telling the sensory system what its last action was. This brain analogy can be extended to the whole organism, in which control is parallel, asynchronous, and spatiotemporally matched with the real world.

A classic way of studying cognitive science was a serial process of "sense, think, and act," which guided many artificial intelligence robot designs. In response to this, Rodney Brooks and Ron Arkin developed behavior-based robots that responded asynchronously to events (Arkin, 1998; Brooks, 1986). Brooks's subsumption architecture is worth examining because it facilitates multitasking and is event driven. Furthermore, it has been suggested that the nervous system operates in a similar fashion to subsumption architectures (Prescott et al., 1999).

5.6.1 Subsumption Architecture

Rodney Brooks proposed the subsumption architecture to show that intelligence could emerge without reason and without representation (Brooks, 1991). This was in stark contrast to the symbolic representation approaches that dominated artificial intelligence at that time. The subsumption architecture was demonstrated on a series of robots, which although simple were reactive, responsive, and reminiscent of real organism behavior.

In the subsumption architecture robot behaviors are arranged in different levels, with the top levels subsuming the levels below. This means that the upper levels can access behaviors from the lower levels and depend on these lower levels to perform their own behaviors. For example, Brooks created the Allen robot following this architecture. Exploring the world is a top-level layer of the robot. This is dependent on the level below, which is to wander around. Wandering around is further dependent on the lower level of avoiding objects, because it does not want to hit any obstacles while wandering. All three of these levels have access to sensor input and operate in parallel. However, only the lowest level communicates with the actuators and results in actual movement from the agent. After the agent actuates a movement, the process loops back to continue taking input from the sensors.

Brooks emphasizes four principles that are supported by subsumption architecture: situatedness, embodiment, intelligence, and emergence.

1. *Situatedness* in robotics means to have a close relationship with the environment, relying more on responding to current environmental situations rather than on long-term, top-down planning.

2. *Embodiment* is the instantiation of an agent in an actual environment, as opposed to a simulation or abstract computational framework. Subsumption architecture describes systems that effectively should be implemented as physical robots in the real world. This results in a robust system that is not biased by the simplicity and predictability of a simulation.

3. *Intelligence* is defined as the ability to perceive and move about in the world. Through this dynamic interaction with the environment intelligent behavior is realized.

4. *Emergence* occurs when seemingly intelligent and unexpected behaviors arise out of simpler behaviors. This is true in subsumption architecture, as the subbehaviors alone do not seem intelligent but combine in a hierarchical fashion to create an intelligent agent.

5.6.2 Layered Hierarchical Control in the Nervous System

There are parallels between the subsumption architecture's layered hierarchical design and the nervous system (see figure 5.8). Neuroscientist Larry Swanson proposed a basic plan for the nervous system that somewhat follows this design; a four-component functional systems model with a motor system controlling behavior and visceral functions (i.e., internal organs and bodily functions), whose output is a function of activity in sensory, cognitive, and behavioral state systems, with feedback throughout (Swanson, 2007). It should be noted that in this view the brain or central nervous system is only one component. Areas that regulate basic behaviors and internal monitoring are subcortical.

Prescott, Redgrave, and Gurney suggested that the hierarchical, layered subsumption architecture could describe the neural mechanisms of defense for the rat (Prescott et al., 1999). Figure 5.9 shows a subsumption architecture diagram for the defense behavior. The

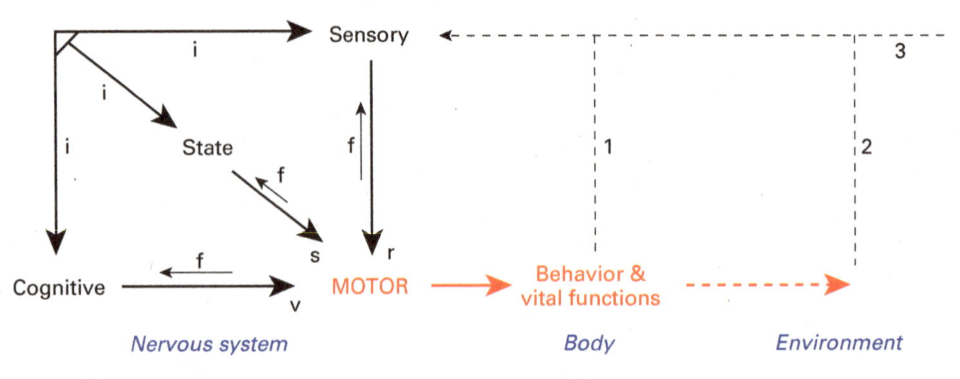

Figure 5.8
Swanson's four-component global model of vertebrate nervous system organization. Behavior is a product of motor subsystem activity, which is a function of activity in three other subsystems: sensory, behavioral state, and cognitive. Cognitive information elaborated by the cortex mediates voluntary control of behavior. Vital functions within the body produce feedback through the sensory subsystem, as do the effects of behavior and vital functions on the environment. From Swanson (2007).

lower levels are reactive and include withdrawal, startling, and freezing. The higher levels can subsume the lower levels by suppressing behavior or predicting outcomes through conditioning. Sensory input provides stimuli that can trigger behavioral responses. Higher levels can set states or context that may shape responses. Note that the diagrams in figures 5.8 and 5.9 follow the *Degeneracy* design principle described previously. There are multiple ways to generate the same motor response; the same architecture can generate multiple responses; and there are multiple ways that a rat can defend itself.

5.6.3 Event-Driven, Parallel Computing

Multitasking and event-driven processing is prevalent in current technology due in part to the ubiquity of real-time and embedded systems. Most current computing devices, from smartphones to desktops, onboard automotive computers to entertainment systems, have parallel processes to handle asynchronous events. Neuromorphic computing architectures developed by researchers and major chip companies such as IBM and Intel are asynchronous and highly parallel, and they are composed of many computing units that act like neurons (Davies et al., 2018; Merolla et al., 2014). This architectural design also follows the *Energy Efficiency through Cheap Design* principle described previously. Neuromorphic hardware architectures use orders of magnitude less power than conventional computing by not relying on a synchronous clock and using spiking elements (Krichmar et al., 2019). A neuron uses most of its energy when it fires an action potential and when an action potential is processed at the synapse. Because neurons do not fire often (in the typical range of 10 to 100 Hz), the nervous system is in low-power mode between spikes. This idea was not lost on most neuromorphic hardware designers. Furthermore,

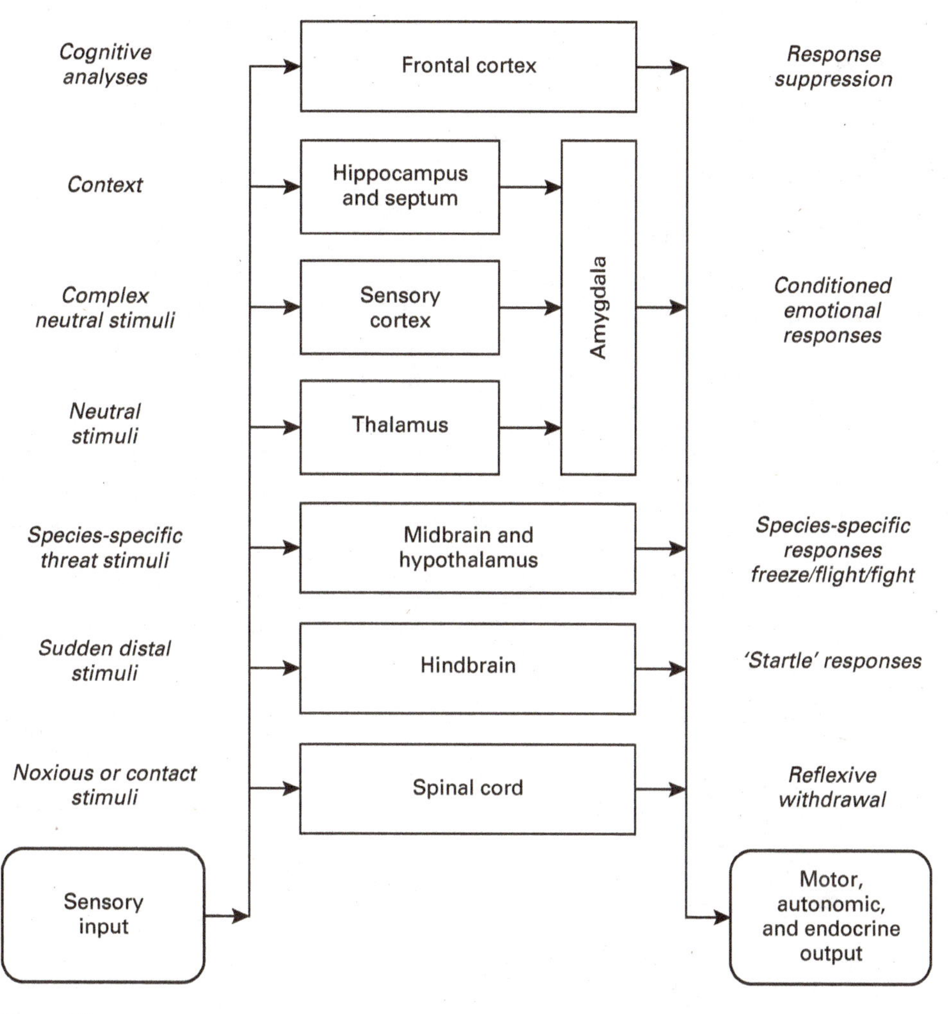

Figure 5.9
Subsumption architecture diagram for neural areas controlling defensive behavior. Higher-level components provide increasingly sophisticated solutions to problems of reducing and avoiding harm. The nature of the sensory input, the principal brain structures involved, and the nature of the defense reaction are indicated for each level. From Prescott et al. (1999).

communication bandwidth is reduced because information is sent only when there is a spike event.

5.7 Case Study: Action Selection in Neurorobotic Model of the Basal Ganglia

A recurring theme in the basal ganglia literature is that these structures operate to release inhibition from desired actions while maintaining or increasing inhibition on undesired actions. The Adaptive Behavior Research Group at the University of Sheffield developed the idea that the basal ganglia acts as an action selection mechanism by resolving conflicts between functional units that are physically separated within the brain but are in competition for behavioral expression (Prescott et al., 2006).

The principal structures of the rodent basal ganglia (see figure 5.10*A*) are the striatum (consisting of the caudate, the putamen, and the ventral striatum); the subthalamic nucleus (STN); the globus pallidus (GP); the substantia nigra (SN, consisting of the pars reticulata SNr and pars compacta SNc); and the entopeduncular nucleus (EP) (homologous to the globus pallidus internal segment [gPi] in primates). These structures are massively interconnected and form a functional subsystem within the wider brain architecture (see figure 5.10*B*).

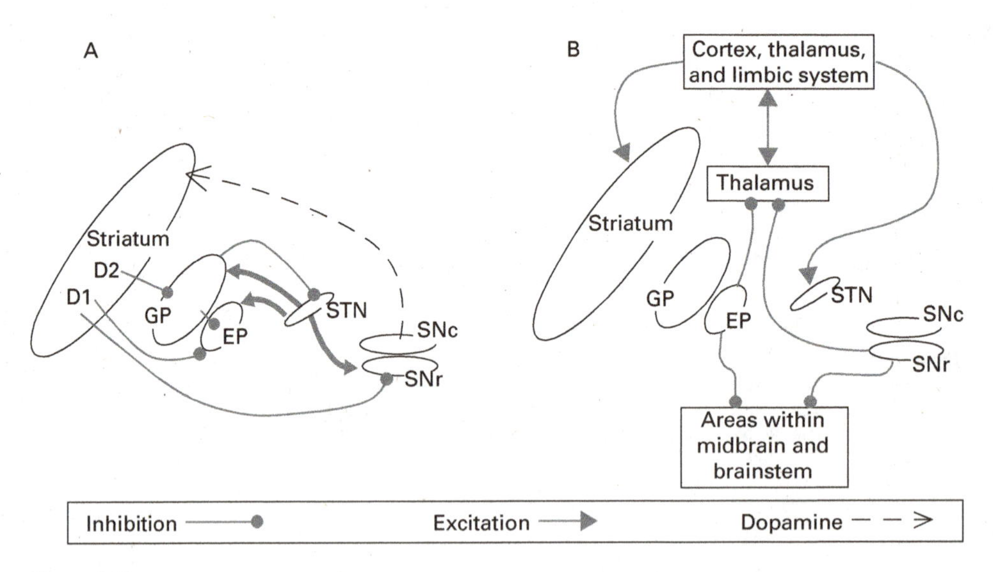

Figure 5.10
Basal ganglia anatomy of the rat. *A*, Internal pathways. *B*, External pathways. Not all connections are shown. Abbreviations: STN, subthalamic nucleus; EP, entopeduncular nucleus; GP, globus pallidus; SNc, substantia nigra pars compacta; SNr, substantia nigra pars reticulata; D1 and D2, striatal neurons preferentially expressing dopamine receptors subtypes D1 and D2. From Prescott et al. (2006).

Prescott and colleagues constructed a system-level model of the basal ganglia and associated thalamocortical connections, which was the control architecture of a small mobile robot engaged in a foraging task. The task required the robot to select appropriate actions under changing sensory and motivational conditions.

A key assumption of their basal ganglia model is that the brain is processing, in parallel, many sensory and cognitive streams or channels, each one potentially carrying a request for an action to be taken. The full robot control architecture is illustrated in figure 5.11. The control architecture contains (1) the robot and its sensory and motor systems; (2) the embedding architecture that provides a repertoire of action (behavioral) subsystems, computes their relative salience, and combines their outputs subject to gating by the basal ganglia; and (3) the extended basal ganglia model that provides the substrate for resolving action selection conflicts. Neuronal unit activity was governed by leaky integrators. Interested readers can find more details on the model in the original model (Gurney et al., 2001a, b; Prescott et al., 2006).

To embody the basal ganglia model, they used a Khepera robot with two wheels and a gripper arm. The robot had eight infrared proximity sensors. The control architecture of

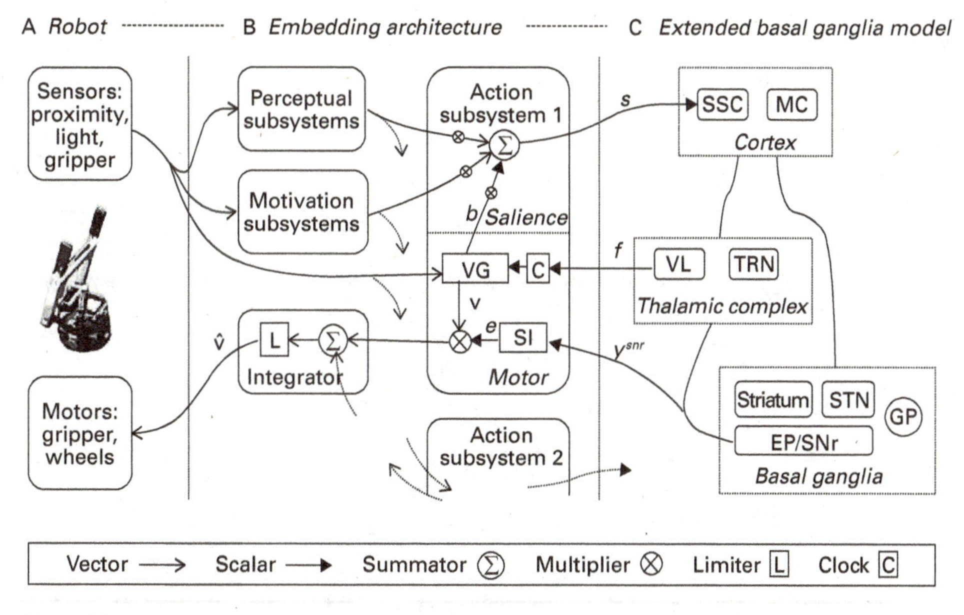

Figure 5.11
The embedded basal ganglia model. The model is composed of three parts: *A*, The robot and its sensory and motor primitives; *B*, The embedding architecture—a repertoire of perceptual, motivational, action (behavioral) subsystems; and its interface to *C*, the biomimetic extended basal ganglia model. Connections for the first of the five action subsystems are shown (projections to and from other action subsystems are indicated by dotted lines). Abbreviations: VG, vector generator; SI, shunting inhibition; b, busy signal; s, salience signal; f, feedback signal; y^{snr}, basal ganglia output; e, gating signal; v, motor vector; $\hat{v}$, aggregate motor vector. From Prescott et al. (2006).

the robot included five behaviors, each corresponding to an action subsystem (see figure 5.12): (1) searching for cylinders, (2) picking up a cylinder, (3) looking for a wall, (4) following along a wall, and (5) depositing the cylinder in a corner. Each action subsystem operated independently to compute a stream of output signals that were directed toward the robot motor systems. So, for instance, cylinder-seek used the infrared proximity sense to detect nearby surfaces and to discriminate objects that were likely to be cylinders from other contours such as walls, and generated motor outputs that specified movement toward or away from the stimulus object as appropriate.

The task for the basal ganglia model was to arbitrate at each time step between the five available action subsystems and to generate a pattern of action selection over time, resulting in coherent sequences of behavior. Two intrinsic drives, analogous to hunger and fear, drove the behavioral choices. Fear was calculated as a function of exposure to the environment and reduced with time spent in the environment; hunger gradually increased with time and was reduced when cylinders were deposited in the nest corners of the arena. Cylinder seeking and cylinder pickup were driven mainly by hunger. Cylinder deposit was driven by proximity to a nest and having a cylinder to deposit. Wall seeking and wall following were driven more by fear.

The robot selected an action by running the model until the basal ganglia activity converged. New sensor data received from the robot provided input to the cortex part of the model. This generated activity in the basal ganglia and the thalamic complex. The most active action subsystem had the smallest output from the basal ganglia (y^{snr} shown in figure 5.11), which released the inhibition from the thalamic complex and gated activity

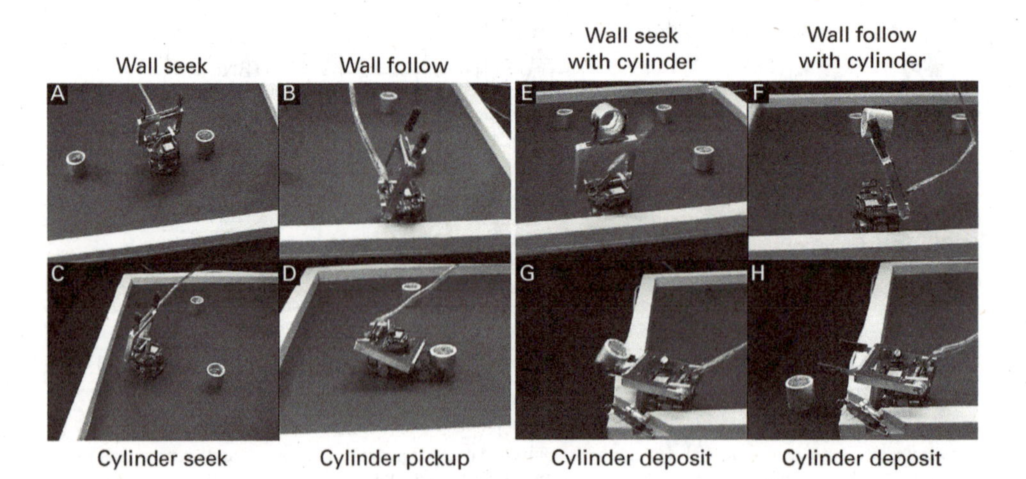

Figure 5.12
Elements of robot behavior in the simulated foraging task. *A*, Wall-seek. *B*, Wall-follow. *C*, Cylinder-seek. *D*, Cylinder-pickup. *E*, Wall-seek (carrying a cylinder). *F*, Wall-follow (carrying a cylinder). *G* and *H*, Cylinder-deposit. From Prescott et al. (2006).

(e shown in figure 5.11) to the robot. Once an action was selected, all other actions were suppressed until the behavioral pattern was complete. Figure 5.13 shows typical behavior during a robot trial.

The embedded basal ganglia model generated sequences of integrated robot behavior. The robot switched cleanly and decisively between successive behaviors, interrupting an ongoing behavior whenever there was a competitor with significantly higher salience. By generating a model whose behavior was directly observable, this neurorobot experiment

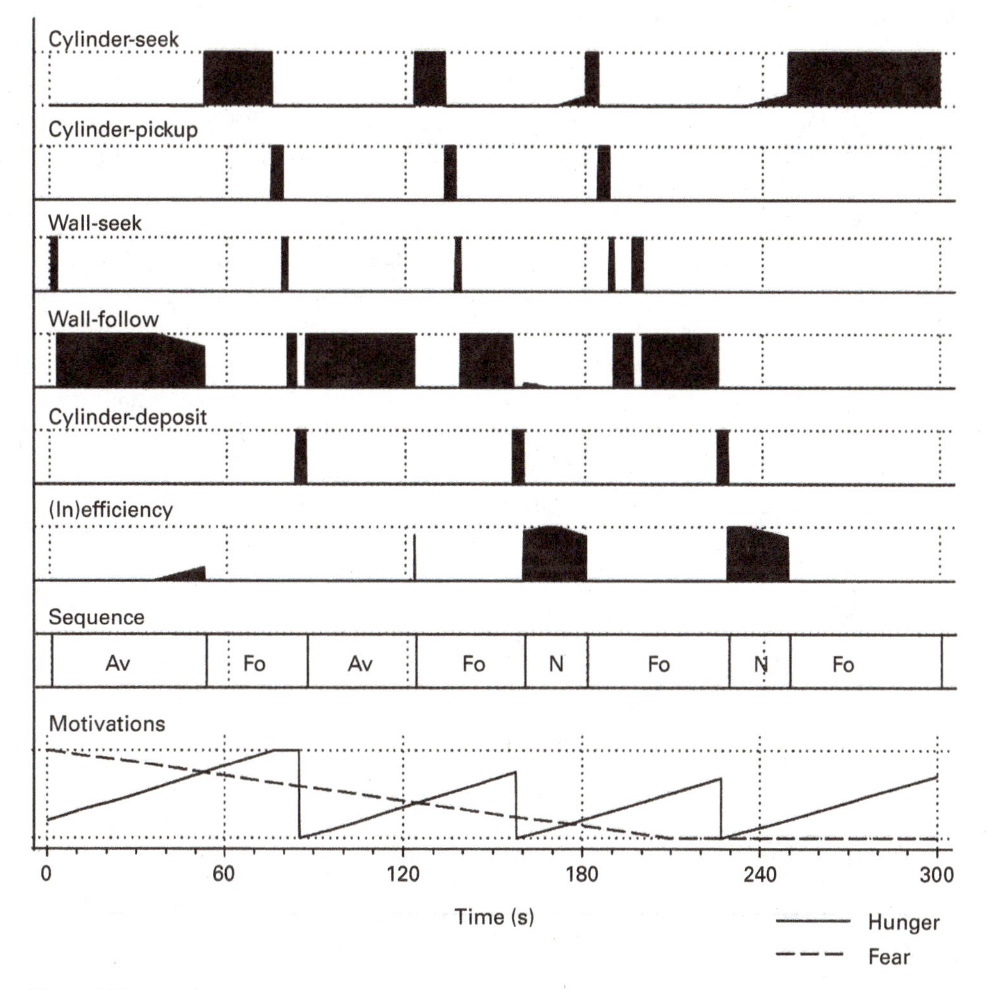

Figure 5.13
Action selection for a 300-second robot trial. From the top down, the first five graphs show the efficiency of selection for a given action subsystem plotted against time; the sixth graph shows the inefficiency of the current winner; the seventh shows which of these actions were avoidance (Av) or foraging (Fo); and the bottom graph shows the levels of the two simulated motivations. From Prescott et al. (2006).

was able to draw interesting comparisons with the outcomes of behavioral experiments with animals: (1) the role of the basal ganglia in behavioral sequencing, (2) the activity of neurons in basal ganglia input (striatum and STN) and output (SNr) nuclei during ongoing behavior, and (3) the behavior of animals in situations of behavioral conflict. This outcome supports the hypothesis that the functional properties of basal ganglia circuitry make it suited to the task of resolving selection conflicts.

Box 5.1 describes a simplified simulation of Prescott and colleagues' neurorobot behavior.

5.8 Summary and Conclusions

This chapter described design principles to consider in creating a neurorobot. In this chapter, we focused on those design principles that react to events with behavior. First and foremost, neurorobotics stresses embodiment. Robot simulators and game engines have come a long way. But ultimately we want these robots engaged in the real world, in which there is reduced chance of bias and oversimplification. Operation in the real world requires energy efficiency as can be observed in the cheap design of the brain and body. The close coupling of the brain, body, and environment requires that sensory information is tightly integrated with motor responses, so much so that the delineation between sensory and motor systems is blurred. Sensory stimuli generate motor responses, and motor responses generate new sensory information. Because there are multiple senses and numerous ways to move the body, degeneracy is a common theme in natural systems. The principle of degeneracy can lead to robustness and flexibility in artificial systems. In conjunction with these principles we need an overall architecture that quickly responds to environmental

Box 5.1

A Webots simulation to examine action selection in a state-event design similar to subsumption architecture can be found in the ActionSelection folder on GitHub: https://github.com/jkrichma/NeurorobotExamples/.

This simulation is a simplified version of the behavior introduced in Prescott et al. (2006). Similar to their study, the Khepera robot decides to eat, hide by a wall, or explore the environment. There is a fear level and hunger level that influences whether to hide or eat. The behavior is dictated by a state-event architecture as shown in figure 5.14. There is a Markov decision process to decide whether to hide or eat when the robot sees an object (i.e., the red cylinder). If the robot hides, it will turn away from the cylinder and hide by a wall. If the robot eats, it will approach, pick up, and put down the red cylinder. The red cylinder will then be placed in a random position in the maze. The fear level decreases with time. The hunger level starts low, increases with time, and then drops to zero when the robot enters the Eat state.

Box 5.1
(continued)

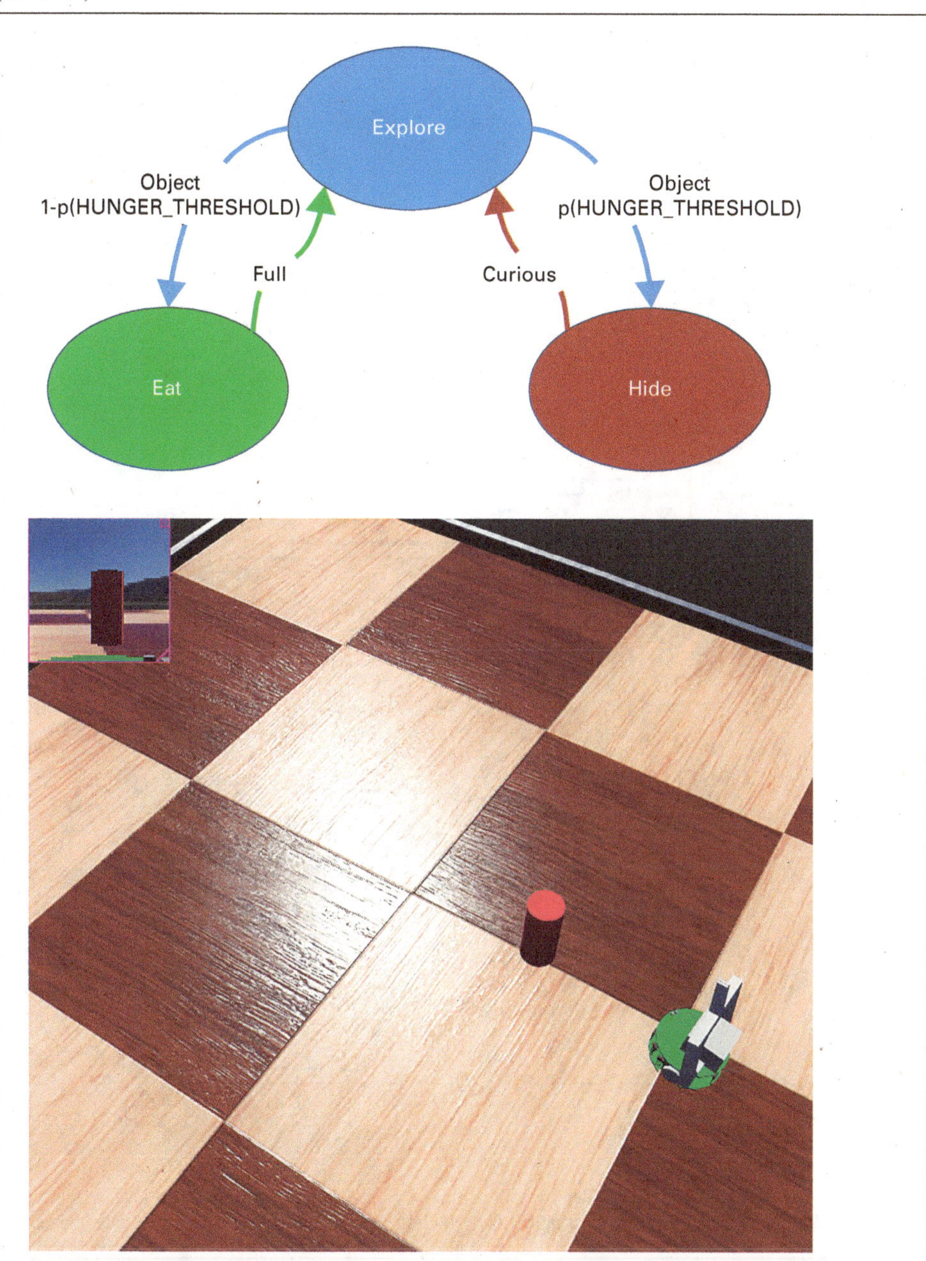

Figure 5.14
Action selection with a Khepera robot. *Top*: The robot has three states (explore, eat, and hide). An object event (seeing a red cylinder) can transition from the explore state to the eat state or the hide state. After eating, which is picking up the red cylinder, the robot is full and transitions to the explore state. After hiding, the robot is curious and transitions to the explore state.

(continued)

Box 5.1
(continued)

The simulation produces a file, called "trial_metrics.txt," that records the robot's state, fear level, hunger level, and position in the arena. These can be examined for analysis. For example, figure 5.15 shows a trial in which the robot hides early in the trial and then explores later.

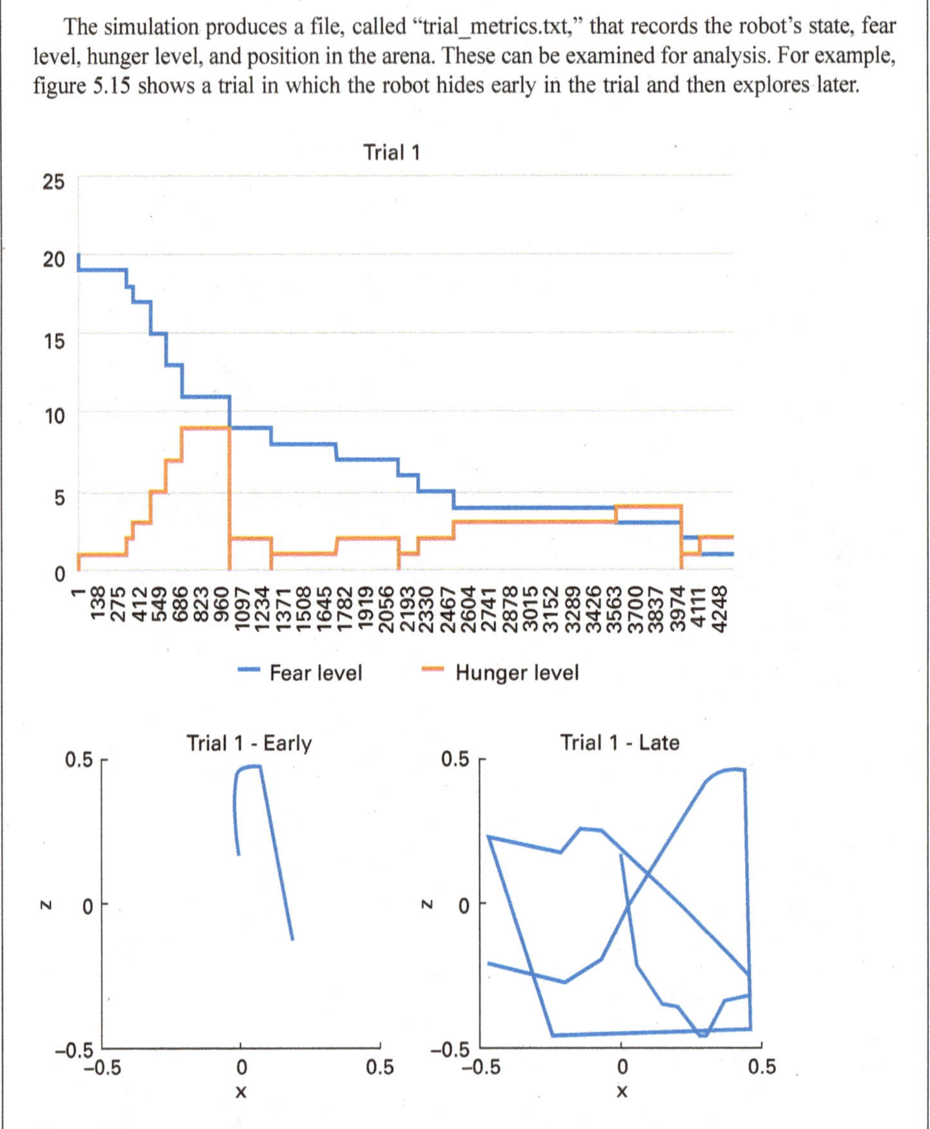

Figure 5.15
Representative trial from the action selection simulation. *Top chart*: The hunger and fear level over time. *Bottom-left chart*: Robot's position during the early portion of trial. *Bottom-right chart*: Robot's position during the late portion of the trial.

The hunger and fear thresholds are static. What happens if they change more dynamically due to the robot's needs? Add another cylinder so that one cylinder is food (prey) and the other cylinder is something scary (predator). Try varying the exploration/exploitation (beta) parameter to see how this affects the robot's behavior.

events with multiple tasks that can operate simultaneously, which leads to parallel, closely coupled architectures such as subsumption and the basic plan of the nervous system.

The design principles discussed in this chapter can be demonstrated in neurorobots with limited capacity to learn and remember. In the next chapter we discuss design principles that endow the neurorobots with learning and memory.

6 Neurorobot Design Principles 2: Adaptive Behavior, a Change for the Better

6.1 Introduction

In this chapter we explore design principles related to adaptive behavior. Adaptation requires learning and remembering what was learned so that it can be applied in the future. Motivation is a key driver of learning. Motivators take many forms, which are called value systems. Another key aspect of adaptive behavior is the ability to predict future events. This requires building up a memory of expectations and the ability to adjust when expectations do not meet the current situation. In chapters 3 and 4 we described learning rules for value-dependent learning and goal-driven behavior. Most of the neurorobots in this chapter use variations of these learning rules to demonstrate adaptive behavior.

6.2 Learning and Memory

As far back as 1950 Alan Turing realized that machines would need to learn in order to show intelligent behavior (Turing, 1950). Although most people recognize Turing for the imitation game and the Turing machine, Turing also suggested that learning was key for creating cognitive machines. He stated that "instead of trying to produce a program to simulate the adult mind, why not rather try to produce one which simulates the child's? If this were then subjected to an appropriate course of education one would obtain the adult brain" (456). This notion led to a subfield of robotics known as cognitive developmental robotics, which is beyond the scope of this discussion; interested readers can find more information in Asada et al. (2009) and Cangelosi and Schlesinger (2015).

In addition to developmental learning, Turing also predicted the field of machine learning: "The machine has to be so constructed that events which shortly preceded the occurrence of a punishment-signal are unlikely to be repeated, whereas a reward-signal increased the probability of repetition of the events which led up to it" (457). These concepts carry over into artificial intelligence and robotics today. Still, the learning and memory ability of the brain is unparalleled in comparison with artificial systems.

Unlike artificial systems, our brains allow us to learn quickly, incrementally, and continuously. With just a few presentations of something new, we can learn to recognize an object or situation or even learn a new skill. When we learn something new, we do not forget what we have learned previously. Moreover, we can take what we learn from one situation and apply it to another. On the other hand, artificial learning systems struggle under these situations, suffer from **catastrophic forgetting** of previous learning when something new is learned, and have difficulties generalizing learning from one task to another.

6.2.1 Memory Consolidation and Spatial Memory

A brain region important for learning and memory is the hippocampus (see figure 6.1). The hippocampus is necessary to learn new memories and to consolidate those new experiences into long-term memories that can last a lifetime. The hippocampus can rapidly learn new autobiographical and semantic information, sometimes in the first experience (i.e., one-shot learning). Over time, this information becomes consolidated in the rest of the brain. Having a rapid learning system that can interact with a slower long-term storage area, which has been called a **complementary learning system** (Kumaran et al., 2016; McClelland et al., 1995), is thought to be the means by which our brains overcome catastrophic forgetting (i.e., forgetting previously learned information when learning new information). This aligns with the another memory model, known as the **hippocampal indexing theory** (Teyler and DiScenna, 1986), which states that memories in the form of neocortical activation patterns are stored as indices in the hippocampus that are later used to aid recall. Although this may be an oversimplification, the notion that the hippocampus and medial

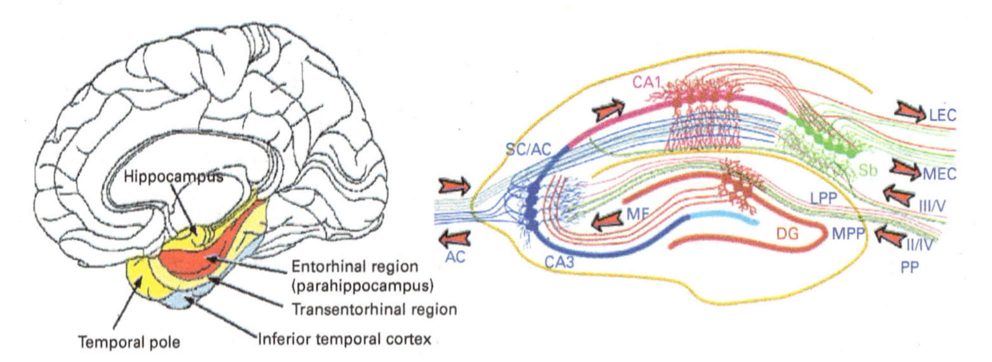

Figure 6.1
The hippocampal formation. *Left*: Medial temporal lobe. The medial temporal lobe, which contains the hippocampus, is important for episodic and semantic memory. Information from many areas of the brain enters the medial temporal lobe. *Right*: Hippocampus connectivity. Cortical information enters the hippocampus through the lateral and medial entorhinal cortex (LEC and MEC). The LEC and MEC mainly project to the dentate gyrus (DG), which then projects to the CA3 subfield and then to the CA1 subfield. From CA1, information gets back to the cortex via the entorhinal cortex.

temporal lobe integrates multimodal information from the neocortex makes sense and is backed by experimental evidence.

In chapter 5, we described the neurorobot Darwin X, which was capable of episodic and spatial memory (Krichmar, Nitz, et al., 2005). Darwin X had an extensive model of the hippocampus and surrounding areas (see figures 6.2*A* and 6.2*B*). The hippocampus, which has different subfields, has strong bidirectional connections with the visual *what* and *where* streams, inferotemporal cortex (IT), and parietal cortex (Pr), respectively. It also gets vestibular self-motion information from the anterior thalamic nucleus (ATN) that can give a signal of heading or head direction. This multimodal sensory information is used to build up memory of places and goals.

In the rodent, spatial memory has traditionally been tested using the Morris water maze, named after memory researcher Richard Morris. In this maze task, the rodent swims in a tank of water until it locates a platform beneath the surface of the water. Rats are good swimmers, but they are strongly motivated to get out of this situation. The water is typically murky so that they cannot see the platform, and there are landmark cues on the walls of the room. If the rat starts each trial from a different location, it cannot use a series of movements to find the platform; rather it must learn the spatial layout of this environment. Over many studies it has been shown that this test of spatial memory requires a hippocampus.

As discussed in section 5.5, the Darwin robot was designed to solve a dry version of the Morris water maze. After roughly the same number of trials it would take a rat to solve this task, Darwin X also demonstrated a memory of the layout of this environment (see figures 6.2*C* and 6.2*D*). In several detailed analyses, it was shown that memories were stored and recalled through a dynamic interaction between the modeled hippocampus and the surrounding cortical areas (Krichmar, Nitz, et al., 2005; Krichmar, Seth, et al., 2005). In particular, the entorhinal cortex (EC_{IN} and EC_{OUT}) played an important role in this dynamic memory. We revisit these brain areas in chapter 8 because they are also important for navigation in animals.

Box 6.1 describes a simulation of the Morris water maze using model-free reinforcement learning and place cells.

6.2.2 Context and Schemas

Our memories have context, and this contextual information can help us generalize when we encounter novel yet similar situations. In the literature this is called a **schema**, which is the memory of a set of items or actions bound together by a common context (van Kesteren et al., 2012). For example, if you are in a restaurant, you expect to see tables, chairs, a menu, waiters, and so forth. If you go to a new restaurant, that common context information can be used to rapidly consolidate the novel information into the restaurant schema. This requires mental representations that are flexible enough to learn tasks in new contexts and yet stable enough to retrieve and maintain tasks in old contexts.

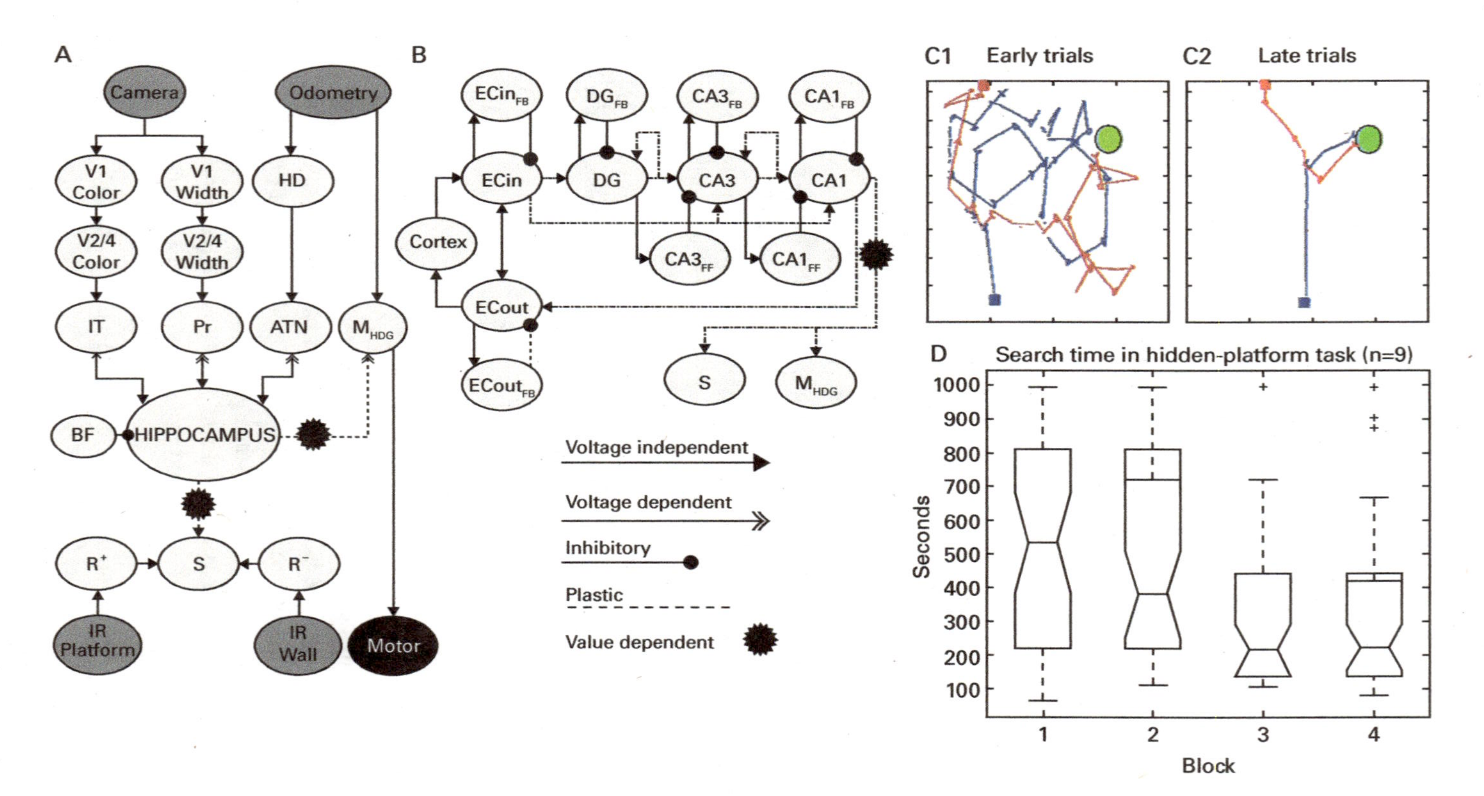

Figure 6.2
Schematic of the neural network model for Darwin X. Gray ellipses denote different neural areas. Arrows denote projections from one area to another. *A*, Diagram of cortical-hippocampal connectivity. Input to the neural simulation comes from a camera, wheel odometry, and IR sensors for wall and platform detection. The simulation contains neural areas analogous to the visual cortex (V1, V2/4), the inferotemporal cortex (IT), the parietal cortex (Pr), head direction units (HD), the anterior thalamic nuclei (ATN), motor areas for egocentric heading (M_{HDG}), a value system (S), and positive and negative reward areas (R+, R−). *B*, Diagram of connectivity within the hippocampal region. The modeled hippocampus contains areas analogous to the entorhinal cortex (EC_{IN}, EC_{OUT}), the dentate gyrus (DG), and the CA3 and CA1 subfields. These areas contain interneurons that implement feedback inhibition (e.g., CA3 → CA3FB → CA3) and feed-forward inhibition (e.g., DG → CA3FF → CA3). *C*, Trajectories of Darwin X. The green circle denotes the location of the platform during training trials. Red and blue squares denote starting locations. Red and blue lines indicate trajectories during individual trials. *C1*, Early trials. *C2*, Late trials. *D*, Search times over the course of the experiment. Each Darwin X subject ran sixteen trials in which a block consisted of four trials, each with a different starting location. The middle line of each box is the median search time, the lower line is the twenty-fifth percentile, and the upper line is the seventy-fifth percentile. From Krichmar, Nitz, et al., (2005).

Box 6.1

A Webots simulation to observe a robot solving the Morris water maze can be found in the Morris_Water_Maze folder on GitHub: https://github.com/jkrichma/NeurorobotExamples/.

This simulation implements a model of a rat solving the Morris water maze using a type of model-free reinforcement learning known as an actor-critic model with temporal difference learning. Model details can be found in Foster et al. (2000). The state is a population of place cells, and the action is to move in one of eight compass headings (see figure 6.3).

Figure 6.3
Simulation of the Morris water maze using the Firebird robot from Nex robotics. The robot's compass is used to move in one of eight headings. Supervisor mode is used to activate a population of place cells and to detect whether the robot reaches the hidden platform, which is depicted by the smaller circle.

The simulation outputs files with the place cell locations (mwm_place.txt), actor weights (mwm_z.txt), and the number of time steps to find the platform (mwm_latency.txt). Included in the GitHub folder is a MATLAB script (mwm_analysis.m) to plot results. The left and middle panels of figure 6.4 show results from one trial, and the right panel of figure 6.4 shows the overall results from ten trials.

(continued)

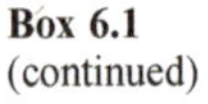

Box 6.1
(continued)

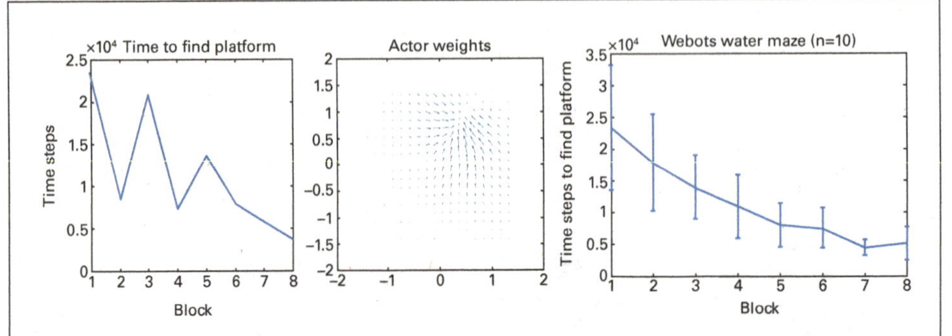

Figure 6.4
The left panel shows the number of time steps to find the platform during a trial. Each block consists of four different starting points. The middle panel shows the actor's weight vectors. The arrows denote the direction and magnitude of eight weights. Following the arrows leads toward the platform. The right panel shows the mean and standard deviations for the time it took to find the platform over ten trials.

The performance of the robot depends on its learning parameters and the place cell responses. Try changing the width or number of place cells. Try changing the learning rate. How do these parameter changes affect learning?

Similarly in neuroscience the stability-plasticity dilemma asks how the brain is plastic enough to acquire new memories quickly and yet stable enough to recall memories over a lifetime (Abraham & Robins, 2005; Mermillod et al., 2013). The adaptive resonance theory (ART) developed by Steve Grossberg and colleagues provides a brain-inspired solution to this problem (Grossberg, 2013, 2017). ART implements the match learning rule to show how top-down expectations focus attention on salient combinations of cues and how attention may operate via a form of biased competition. When a good enough match occurs, a synchronous resonance occurs that increases attentional focus and can drive fast learning of bottom-up recognition categories and top-down expectations. Grossberg proposed that good top-down matches may cause fast gamma oscillations that support attention, resonance, and learning, whereas bad mismatches inhibit learning by causing slower beta oscillations while triggering attentional reset and hypothesis testing.

These predictions for solving the stability-plasticity dilemma fit nicely with experiments showing that rapid consolidation of information occurs when it fits within a schema. Tse and colleagues demonstrated this by training rats on different schemas, which were collections of associations between different foods and their locations in a square arena (Tse et al., 2007). They found that the rats were able to learn new information quickly if it fit within a familiar schema. Additionally, the rats were able to learn new schemas without forgetting previous ones. The hippocampus was necessary for learning schemas and any new information matching a schema. A subsequent study showed increased plasticity in the medial prefrontal cortex (mPFC) when information was consistent with a familiar schema (Tse et al., 2011). However, the hippocampus was not necessary to recall these memories, even after a short time period (e.g., 48 hours). This challenged the idea of complementary learning systems because new information could rapidly be consolidated in cortical memory under these conditions. Our group simulated the schema experiment with a neural network model of the hippocampus and mPFC (Hwu and Krichmar, 2020). Interestingly, we used a contrastive Hebbian learning rule, in which the oscillatory epochs to assimilate new information had similarities to adaptive resonance. In the section 6.5 case study, we describe a neurorobotic version of the model.

6.2.3 Episodic Memory and Schemas in Neurorobots

Endowing robots with memories of events and contexts has practical value, especially in scenarios in which robots are interacting with humans. Several groups have taken inspiration from the neuroscience of episodic memory and schemas to create practical robotic applications. For instance, Sigalas and colleagues implemented hierarchical hidden Markov models (HMMs) to create schemas for a robot's episodic memory (Sigalas et al., 2017). They used a robotic arm to learn a *serving breakfast* scenario that adapted to the user's needs. Zender and colleagues created a robot that built an internal representation through its exploration of an indoor environment (Zender et al., 2008). It built schematic concepts into this map such that a living room has a television and a couch and that a kitchen has a refrigerator and an oven. Similarly, Kostavelis and Gasteratos constructed a robot simultaneous localization and mapping (SLAM) system that added semantic information to its map while it navigated an office space (Kostavelis & Gasteratos, 2017). Although not strictly tied to known properties of brain connectivity and functionality, they are inspired by neurobiological memory systems and demonstrate how this inspiration can lead to natural behavior in human environments.

6.3 Value

Robots should be equipped with a value system that constitutes a basic assumption of what is good and bad for its health and well-being. In the brain, value systems are often equated with neuromodulators or neurohormones. A value system facilitates the capacity of a biological brain to increase the likelihood of neural responses to an external phenomenon (Merrick, 2017). The combined effects of perception, experience, reasoning, and decision making contribute to the development of values in animals. Value can also be thought of as a measure of the effort one is willing to expend to obtain reward or to avoid punishment.

In addition to rewards and punishment, **intrinsic motivation** can be a form of value (Oudeyer and Kaplan, 2007). This can take the form of seeking novelty, fun, play or curiosity, that is, obtaining value for its own sake rather than to satisfy some need. For example, Oudeyer and colleagues created a robotic playground where a Sony AIBO dog with a motivation to explore and learn new things manipulated objects (Oudeyer et al., 2007). The robot first spent time in situations that were easy to learn and then shifted its attention progressively to more difficult situations, avoiding situations in which nothing could be learned.

Using "critters" constructed with LEGO Mindstorms, Merrick showed how value systems, driven by Q-learning in multiple neural network models, led to motivated behavior such as novelty seeking, interest, competence, and random exploration (Merrick, 2010a; Merrick, 2010b). In general, intrinsic motivation is an important topic in machine learning. It allows the system to find solutions in situations in which rewards are few and far between.

For neurorobotics, a robot that can maximize positive and minimize negative values as well as learn for its own sake is important and can lead to interesting behavior and continual learning.

6.3.1 Neuromodulation

Neuromodulators are thought to be the brain's value systems (Friston et al., 1994; Krichmar, 2008). Neuromodulators are chemicals that signal important environmental or internal events. They cause organisms to adapt their behavior through long-lasting signals to broad areas of the nervous system. Neuromodulators in the brain influence synaptic change (i.e., learning and memory) to satisfy global needs according to value.

To shape behavior, cognitive robots should have an innate value system to tell the robot that something was of value and trigger the appropriate reflexive behavior. From this experience the agent can learn which stimuli were predictive of that value and try to maximize the acquisition of good value while minimizing the acquisition of bad value. Many of these value-based robots employ models of **dopamine** to shape behavior (Chou et al., 2015; Dominey, 2013; Fiore et al., 2014; Sporns & Alexander, 2002).

Over the last few years the simulation of value systems has been expanded to include multiple neuromodulatory systems found in the vertebrate (Avery & Krichmar, 2017; Krichmar, 2008). Besides the dopaminergic reward system there is the neuromodulator **serotonin** that is involved in harm aversion or impulsiveness. The noradrenergic system signals oddball or unexpected events. The cholinergic system is thought to increase attention to important features and at the same time to decrease the allocation of attention to distractors. **Acetylcholine** and **noradrenaline** could be thought to signal intrinsic value by allocating attention and triggering learning. These systems interact with each other through direct and indirect pathways, and they all respond strongly to novelty by sending broad signals to large areas of the brain to cause a change in network dynamics resulting in decisive action.

Introducing saliency into the environment can lead to attentional signaling. For example, the robot CARL was designed to test a computational framework for applying neuromodulatory systems to the control of autonomous robots (Cox & Krichmar, 2009). The framework was based on the following premises (Krichmar, 2008): (1) the common effect of the neuromodulatory systems is to drive an organism to be decisive when environmental conditions call for such actions and allow the organism to be more exploratory when there are no pressing events; and (2) the main difference between neuromodulatory systems is the environmental stimuli that activate them. In the experiment, two out of four objects were salient, and CARL learned the appropriate action for each (see figure 6.5). Unexpectedly, a strong attentional bias toward salient objects, along with ignoring the irrelevant objects, emerged through its experience in the real world. The selective attention could be observed both in CARL's behavior and in CARL's simulated brain. These neurorobotic experiments showed how phasic neuromodulation could rapidly

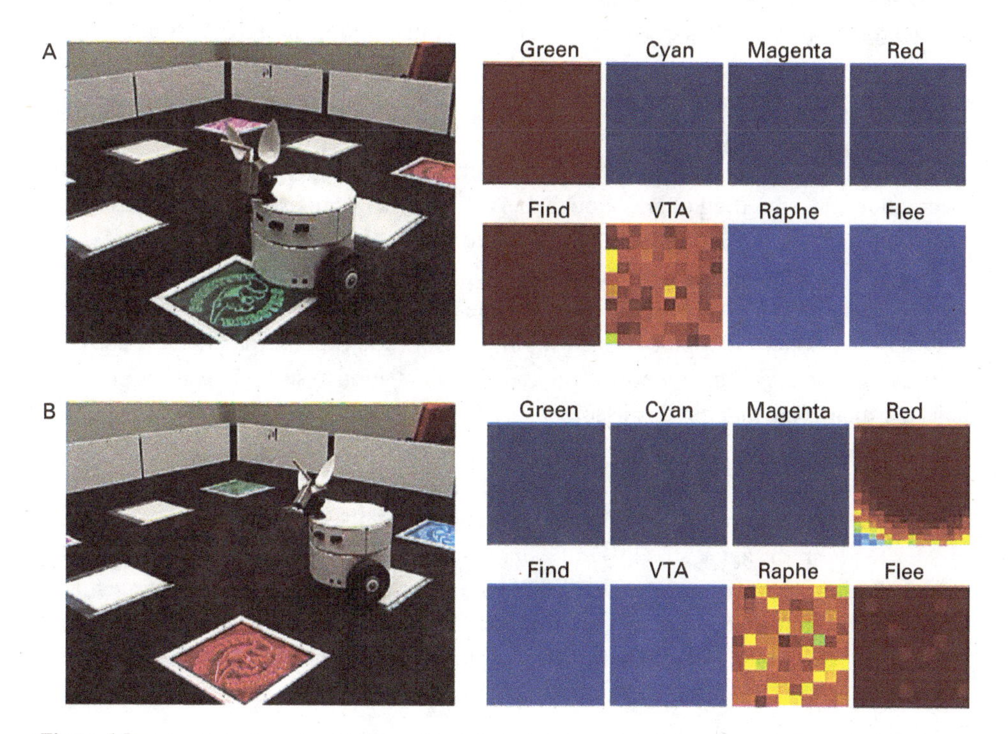

Figure 6.5
CARL robot in colored panel task. The panels could flash any of six different colors. One color, green, signaled positive value. Another color, red, signaled negative value. The remaining colors were neutral. Positive and negative signals were transmitted from the panel to a receiver on the bottom of CARL. *A*, CARL during an approach or *find* response. The panels on the right show strong neuronal activity in its simulated green visual area, the dopaminergic system (VTA), and the *find* motor neurons. *B*, CARL during a withdrawal or *flee* response. The panels on the right show strong neuronal activity in its simulated red visual area, the serotonergic system (raphe), and the *flee* motor neurons.

focus attention on important objects in the environment by increasing the **signal-to-noise ratio** (SNR) of neuronal responses. The model further suggested that phasic neuromodulation amplifies sensory input and increases competition in the neural network by gating inhibition.

The neuromodulatory system also regulates attention allocation and response to unexpected events. Using the Toyota Human Support Robot (Yamamoto et al., 2018), the influence of the cholinergic (ACh) system and noradrenergic (NE) systems on goal-directed perception was studied in an action-based attention task (Zou et al., 2020). In this experiment, a robot was required to attend to goal-related objects (the ACh system) and adjust to the change of goals in an uncertain domain (the NE system). Four different actions (i.e., *eat*, *work-on-computer*, *read*, and *say-hi*) were possible in the experiment and each of them was associated with different images of objects. For example, the goal action *eat* might result in attention to objects such as *apple* or *banana*, whereas the action *say-hi* should increase attention to a *person*. During the experiment, the goal action changed periodically and the robot needed to select the action and object that it thought the user wanted on the basis of prior experience. The cholinergic system tracked the **expected uncertainty** about which goal was valid, and the noradrenergic system signaled **unexpected uncertainty** when goals suddenly changed (Yu & Dayan, 2005). High cholinergic activity levels allocated attention to different goals. Phasic noradrenergic responses caused a rapid shift in attention and a resetting of prior goal beliefs. The model demonstrated how neuromodulatory systems can facilitate rapid adaptation to change in uncertain environments. The goal-directed perception was realized through the allocation of the robot's attention to the desired action/object pair. Figure 6.6 shows the robot deciding which object to bring to the user. The bottom of figure 6.6 shows views from the robot's camera as it correctly guesses that the user's goal is to eat. Its top-down attention system finds an appropriate object, which is an apple in this case.

6.3.2 Real Value versus Simulated Value

One problem that remains unsolved in neurorobotics is that these artificial value systems are dissociated from the agent's body. Real pain, hunger, thirst, and fatigue drive a true value system. Without this connection to bodily dependent drives, an artificial value system does not signal the immediacy of the agent's need and lacks to some degree the ability to shape the agent's behavior. It would be interesting to tie something like the robot's battery level to its *hunger* value. With faster-charging batteries or better solar cells this might be possible.

An interesting step in this direction is work on adaptive robots that can recognize drops in their performance (Cully et al., 2015). Cully and colleagues developed a novel method for adapting gaits on a hexapod robot (see figure 6.7). In a sense, the robot controller had a memory of potential gaits. If one or more of the robot's legs were damaged, the

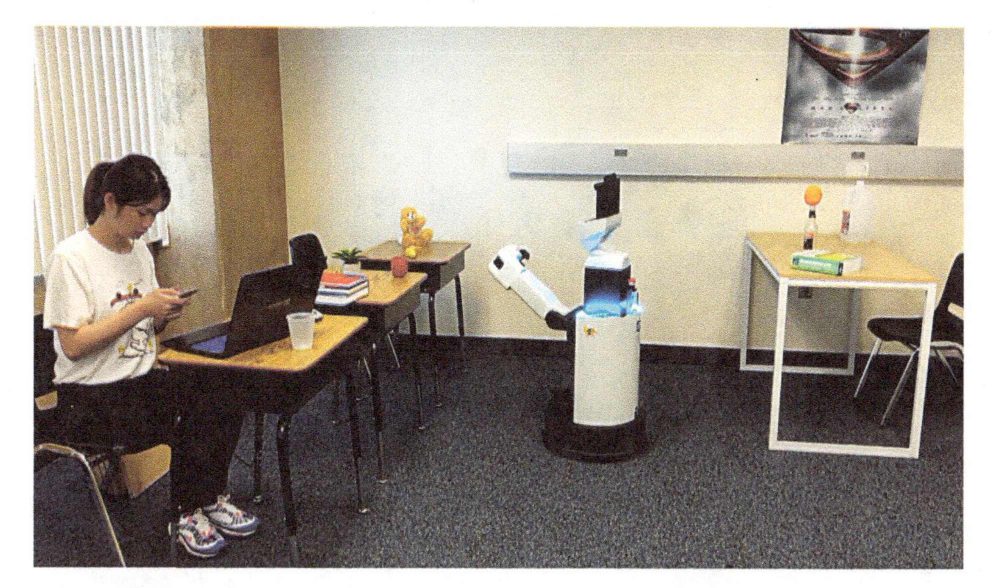

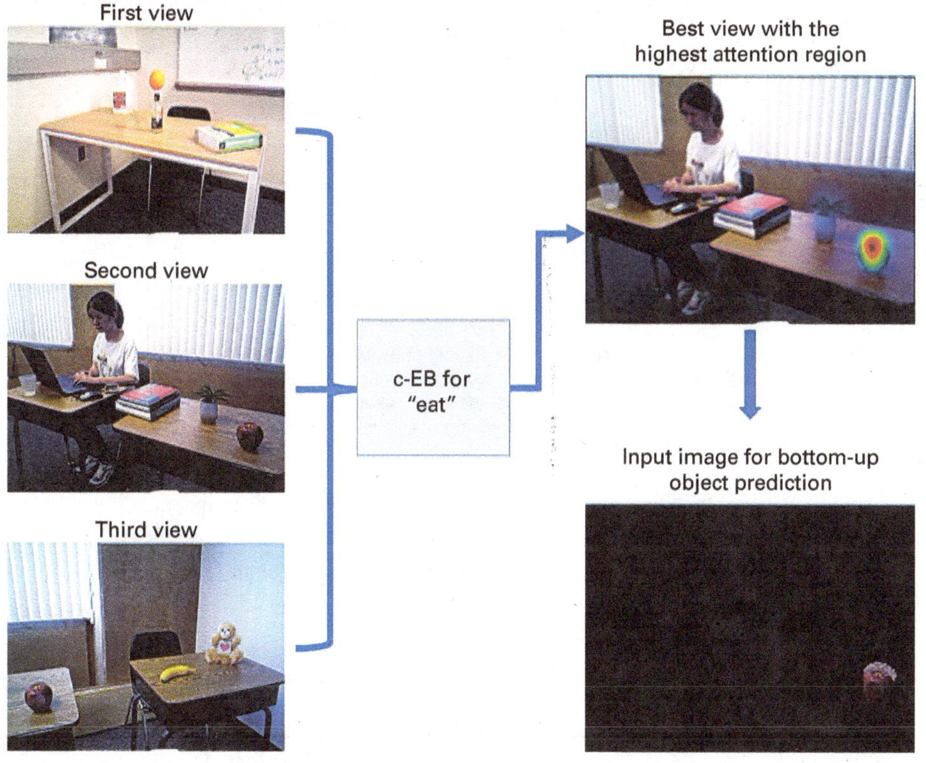

Figure 6.6
Toyota Human Support Robot implementation for the goal-driven perception model, including the top-down attentional search process for a guessed action "eat" based on three different real indoor views, to select the highest attention region for bottom-up object prediction. Adapted from Zou et al. (2020).

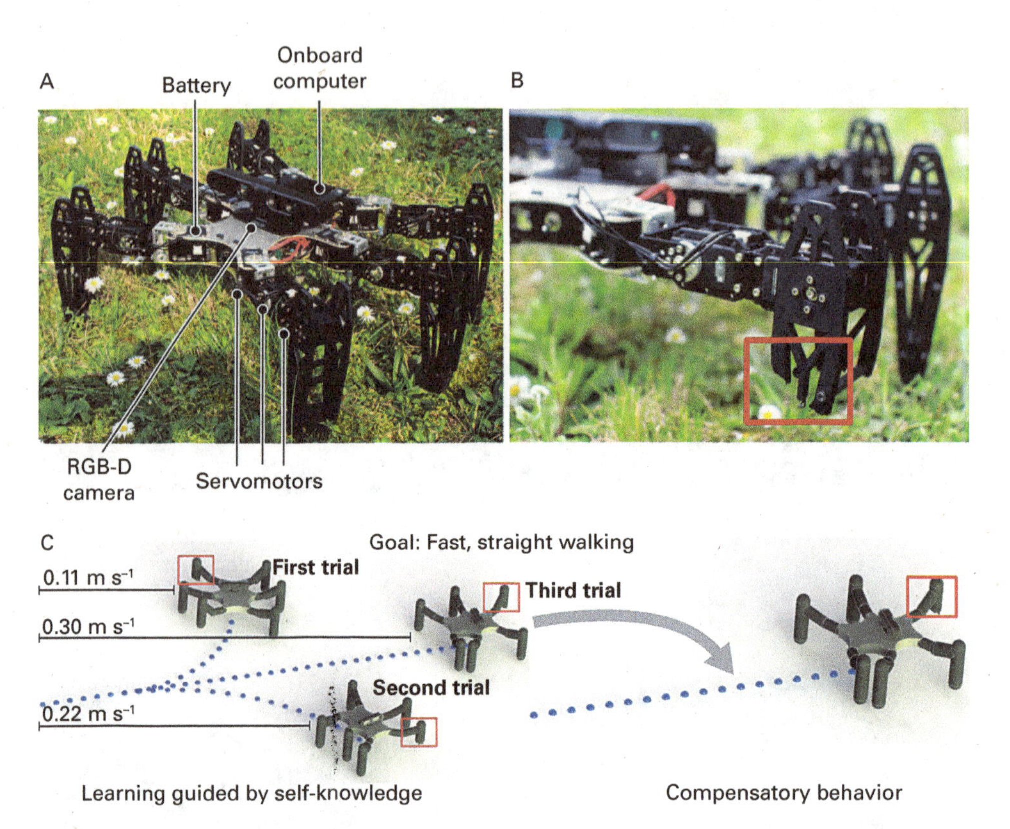

Figure 6.7
Using an imagined trial and error algorithm, robots, like animals, can quickly adapt to recover from damage. *A*, The undamaged, hexapod robot. RGB-D stands for red, green, blue, and depth. *B*, Hexapod robot with a broken leg. *C*, After damage occurs, the robot recognizes it cannot walk fast and in a straight line. The robot tests different types of behaviors until it discovers an effective compensatory behavior. Adapted from Cully et al. (2015).

robot would detect the damage, "imagine" different ways of moving, and then choose the new gait that it thought would work best under the new circumstances. In this way, the robot knew something was wrong and was able to adapt its behavior quickly without intervention.

6.4 Prediction

Predicting outcomes and planning for the future is a hallmark of cognitive behavior. Much of the cortex is devoted to predicting what we will sense or the outcome of a movement or what series of actions will lead to big payoffs. Thus a neurorobot should strive to have these predictive capabilities. Organisms minimize surprises by predicting future outcomes so that they minimize the energy expenditures required to deal with unanticipated events (Friston, 2010). The idea of minimizing free energy has close ties to many existing brain theories, such as Bayesian brain, predictive coding, cell assemblies, and Infomax, as well as Neural Darwinism or neuronal group selection. The notion of surprise can be thought of as the unexpected uncertainty of an event (Yu & Dayan, 2005). In the theory of neuronal group selection (Edelman, 1993), plasticity is modulated by value. Value systems control which neuronal groups are selected and which actions lead to evolutionary fitness; that is, they predict outcomes that lead to positive value and avoid negative value. In this sense predicting value is inversely proportional to surprise.

Prediction is crucial for fitness in a complex world and a fundamental computation in cortical systems (Clark, 2013; George & Hawkins, 2009; Richert et al., 2016). It requires the construction and maintenance of an internal model. Model-based reinforcement learning algorithms require the presence of an internal model and the ability to represent transitions between states (Solway & Botvinick, 2012).

In neurorobotics uncertainty due to sensor noise and environmental change leads to surprise. In neuroscience this implies that the brain constructs an internal model to make predictions about sensory input and action outcomes, thus minimizing surprise, and adapting when these predictions are erroneous. Similarly, as discussed in chapter 4, model-based reinforcement learning constructs a model of transitions between states and the expected outcomes when reaching a state. Biological organisms and robots evolve a policy that minimizes surprise by minimizing the difference between likely and desired outcomes, which involves both pursuing the goal state that has the highest expected utility (exploitation) and visiting a number of different goal states (exploration). In this way novelty seeking and curiosity reduces surprises in the long run (Schwartenbeck et al., 2013).

Prediction has been useful in developing robot controllers. For example, in a humanoid robot experiment it was shown that having a predictive model helped the robot make appropriate reactive and proactive arm gestures (Murata et al., 2014). In the proactive mode the robot's actions were generated on the basis of top-down intentions to achieve intended goals. In the reactive mode the robot's actions were generated by bottom-up sensory inputs in unpredictable situations. In another robot experiment the combination of model-based and model-free reinforcement learning was used in a sorting task (Renaudo et al., 2015). The robot had to push cubes on a conveyor belt. The model-based system improved performance by maintaining a plan from one decision to the next. However, the experiments suggested that the model-free system scales better under particular conditions and may be better in the face of uncertainty. Interestingly, neural correlates of both model-based reinforcement learning and model-free reinforcement learning have been observed in the brain (Daw et al., 2011; Glascher et al., 2010). In general, the brain maintains internal models for a wide range of behaviors, from motor control to language processing (Hickok et al., 2011; Shadmehr & Krakauer, 2008; Shadmehr et al., 2010).

Prediction and internal models also facilitate **mental simulation**. Hesslow proposed a simulation theory suggesting that thinking involves the same processes as interaction with the external environment, but the actions are covert or suppressed (Hesslow, 2012). The theory has the following three components: (1) Simulation of behavior. The frontal cortex is activated in simulating movement, but the movement itself is not carried out. (2) Simulation of perception. During imagination, the sensory cortex is activated as though perceiving stimuli. (3) Anticipation. This enables the sensory motor activity that would have been active if the action had been carried out. This last component ties into our discussion on prediction. This simulated prediction also allows for generating a prediction error if the action does not go as planned. Hesslow and colleagues tested their simulation theory in a simple robot task (Ziemke et al., 2005). They used evolutionary algorithms to evolve a neural network controller to traverse a maze that had distal landmarks. The neural network had a sensorimotor module and a prediction module. When the robot was "blindfolded" by suppressing sensory input, the robot was still able to traverse the maze through simulation via its prediction module.

Jun Tani's group has developed several predictive robot controllers using **recurrent neural networks** (Hwang et al., 2020; Tani, 2016). For example, they trained a hierarchy of continuous time recurrent neural networks (CTRNN) to learn different movements (Yamashita and Tani, 2008). Learning was achieved via backpropagation through time (BPTT), a means to apply an error signal to a sequence of neural activities. A teacher guided the humanoid Sony QRIO robot through different behavioral tasks (see figure 6.8*A*). The CTRNNs received visual information and proprioceptive joint angles from the humanoid robot. Important to the learning were the different timescales of the CTRNNs (see figure 6.8*B*). Slower higher-level CTRNNs sent predictions to the faster lower-level CTRNN units. Prediction errors from the lower levels were propagated to the higher levels for adjustments. Movements that appeared repeatedly were segmented into behavioral primitives. These primitives were represented in fast context dynamics in a form that was generalized across object locations. On the other hand, the slow context units appeared to be more abstract in nature, representing sequences of primitives in a way that was independent of the object location. Tani speculated that this prediction multiple timescale hierarchy had similarities to the cortex (Tani, 2016). Fast responding motor primitives can be found in the primary motor cortex, and the slower prefrontal cortex sends top-down predictions to the primary motor cortex. Similarly, the primary visual cortex sends sensory information and prediction errors to the slower parietal cortex, which sends top-down predictions for the primary visual cortex.

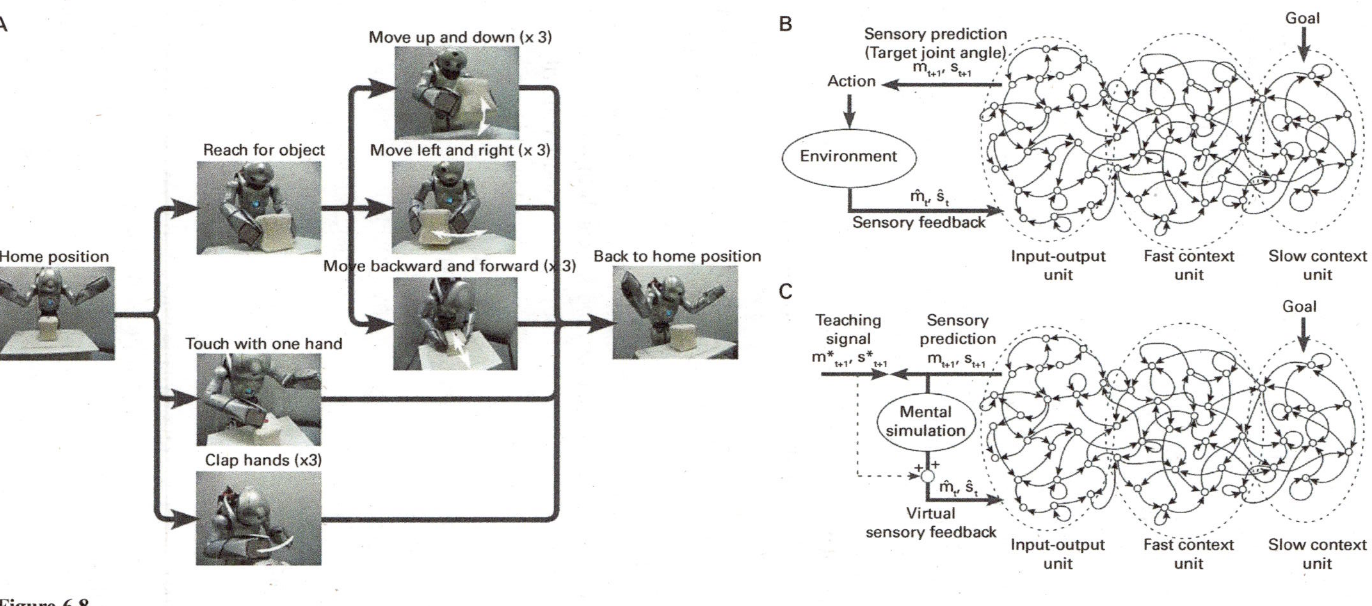

Figure 6.8
A, A humanoid robot was placed in front of a workbench and an object. The task was for the robot to start at a home position, generate one of five different movement behaviors, and then return to the home position. *B* and *C*, Movement behaviors were learned and generated by a hierarchy of continuous time recurrent neural networks (CTRNNs). CTRNNs received sensory input from the robot's camera and joint angle sensors. *B*, Action generation mode. Inputs to the system generated predictions of proprioception and vision. This prediction was sent to the robot as target joint angle motor commands. *C*, Training mode. The network generates behavior sequences and then synaptic weights are updated based on the error between generated predictions and the teaching signals. From Yamashita and Tani (2008).

6.5 Case Study: Schemas and Memory Consolidation in Robots

In section 6.2.2 we discussed how the brain might resolve the stability-plasticity dilemma by grouping contextually related memories into schemas. In a set of experiments with rats the creation and recall of schemas was tested by pairing odors with locations (Tse et al., 2007). When there was a novel context (i.e., a new layout of odors and locations), a schema was created. The researchers went on to show that if a new odor fit within an existing schema it could be rapidly consolidated into the cortex and that the medial prefrontal cortex (mPFC) was necessary to rapidly consolidate information. The hippocampus was necessary for both creating a new schema and updating an existing schema.

In this case study we show how a model of schema and memory consolidation could be applied to a robot task (Hwu et al., 2020). As in the rodent study a schema consisted of a layout of objects and their locations. In this case, it was rooms that one might find in an office or school. Instead of prompting the rat with an odor, the robot was prompted with an object to retrieve. The robot used an object recognition neural network to find objects and a SLAM algorithm to localize itself in the rooms.

A neural network based on interactions between the hippocampus and the mPFC was created to test the ability of the robot to create and utilize a schema (see figure 6.9). A contextual pattern of objects and locations projected to the mPFC, in which each individual neuron encoded a different schema. The ventral HPC (vHPC) and dorsal HPC (dHPC) created triplet indices of schema, object, and location. The vHPC and dHPC drove activity that eventually activated a place cell through a winner-take-all process in the action network. This action neuron caused the robot to go to that location in search of the cued object.

The model contained neuromodulators to encode novelty and familiarity. For example, if an object is novel and the context is unfamiliar, a new schema must be learned. However, if an object is novel and the context is familiar, the object can be added to an existing schema.

Contrastive Hebbian learning (CHL), a biologically plausible learning rule, was used to learn tasks (Movellan, 1991). In general, CHL can be used in a supervised or semisupervised manner to learn associations (see figure 6.10). CHL has a free phase and a clamped phase. In the free phase the input layer is fixed for T time steps until activity converges. Then in the clamped phase, the output is fixed to the desired value to associate with the input, while the input remains fixed. In CHL the desired value for clamping the output can occur in a supervised fashion.

In the case of this experiment the desired value was based on dHPC of the indexing stream (figure 6.9, blue), and the value was learned in the AC and action layers of the representation stream (figure 6.9, orange). The number of clamping and unclamping epochs was related to the novelty and familiarity of an object. It suggests that learning increases when an object is consistent with a schema (i.e., novel object but familiar context). There is evidence that resonance between the mPFC and HPC increases for schema-consistent information (van Kesteren et al., 2012).

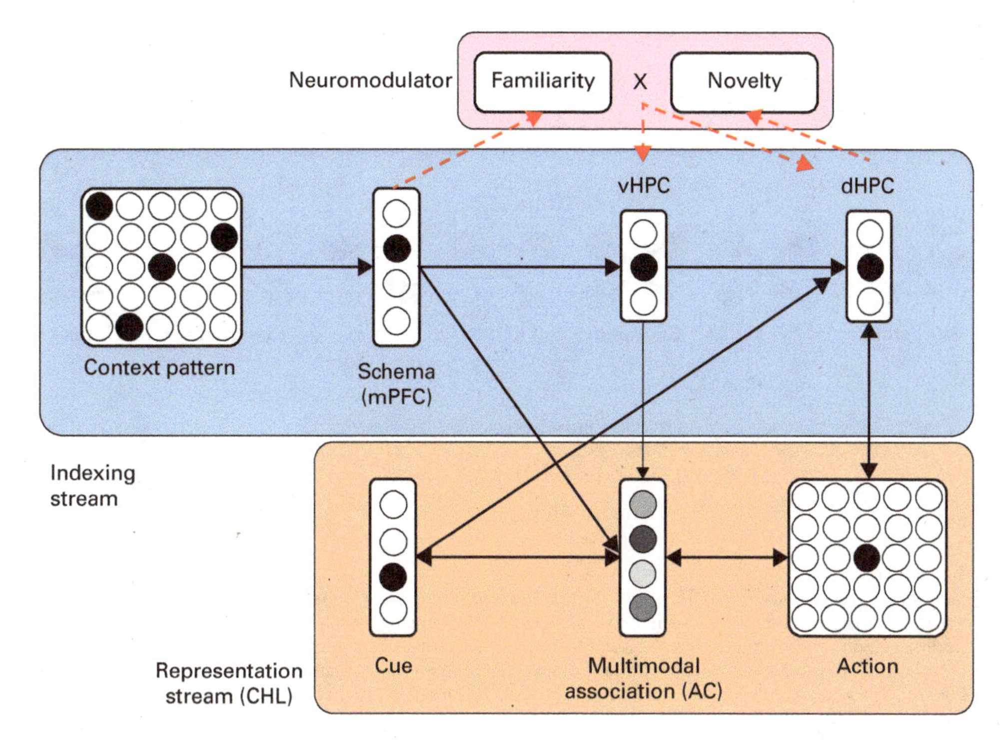

Figure 6.9
Overview of the neural network for schemas. The light blue box contains the indexing stream, including the ventral hippocampus (vHPC) and dorsal hippocampus (dHPC). The light orange box contains the representation stream, including the cue, medial prefrontal cortex (mPFC), multimodal layer (AC), and action layer. Bidirectional weights between layers in the representational stream are learned via contrastive Hebbian learning (CHL). Weights from the indexing stream are trained using the standard Hebbian learning rule. Dotted lines indicate influences of the neuromodulatory area, which contains neurons tracking novelty and familiarity. Neuromodulator activity impacts how often the vHPC and dHPC are clamped and unclamped during learning the task via CHL. From Hwu and Krichmar (2020).

The schema model was embedded on the Human Support Robot (HSR) from Toyota (Yamamoto et al., 2018) and given the task of finding and retrieving objects in a classroom schema and a break room schema (see figure 6.11). In a trial, the robot was prompted to retrieve an object, which required prior knowledge of the schema to which the object belonged and the location of the object. This caused the robot to navigate toward a location, recognize the object, grasp the object, and then return the object to its starting location.

The performance of the robot was tested by typing the name of a graspable object into a graphical user interface and requesting the robot to retrieve the item. The neural network then received input of all objects seen in the last 60 seconds into the context pattern and the requested object as the cue.

The HSR was first placed in a classroom with typical classroom items, shown in figures 6.11*A–B*. The graspable items were an apple, a bottle, and a teddy bear, placed in

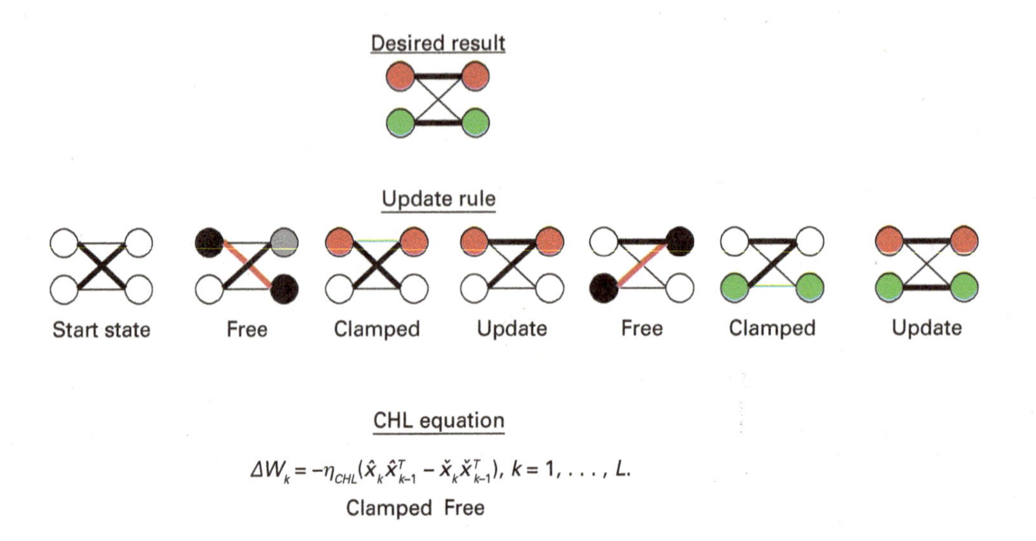

Figure 6.10
Contrastive Hebbian learning (CHL). In CHL, a desired pattern (*top*) is learned by a series of epochs in which network activity has a free phase of activity, followed by the clamping phase, in which the output is fixed to the desired activity, and then updating the weights (*center*). Note that one epoch includes clamping and updating each layer. CHL can iterate a number of these epochs until there is convergence. The bottom section shows the update weight equation for layer *k* of a neural network. For more details see Hwu et al. (2020) and Hwu and Krichmar (2020).

different locations around the room. Figure 6.12*A* shows the performance of the HSR in the classroom. The HSR's search time for objects steadily decreases as it gets more familiar with the classroom schema. After training and testing in the classroom (Exp 1a in figure 6.12*A*), the teddy bear was replaced by a computer mouse (Exp 1b in figure 6.12*A*). The HSR then went through one trial of training and was tested on its ability to retrieve this novel object. Although the computer mouse was a novel object, the HSR knew it belonged in the classroom and was able to consolidate this new information into the existing classroom schema.

Next, the HSR was placed in a second schema consisting of a break room, shown in figures 6.11*C–D*. The graspable items were an apple, a cup, and a wine glass. After training and testing, the classroom schema was tested again to see if the robot was able to maintain performance of prior tasks. As with the classroom, figure 6.12*B* shows that the robot was able to learn this new break room schema without forgetting objects' locations in the classroom (CR in figure 6.12*B*).

Last, the HSR was tested to see whether schemas could help the robot retrieve items that it was never explicitly trained to retrieve. For example, the HSR was cued by showing it a banana. Because the banana was part of the break room schema, the HSR searched for the banana in the break room first rather than in the classroom. Likewise, if the HSR was cued with a book, it searched for the book on the desk in the classroom. The top images in

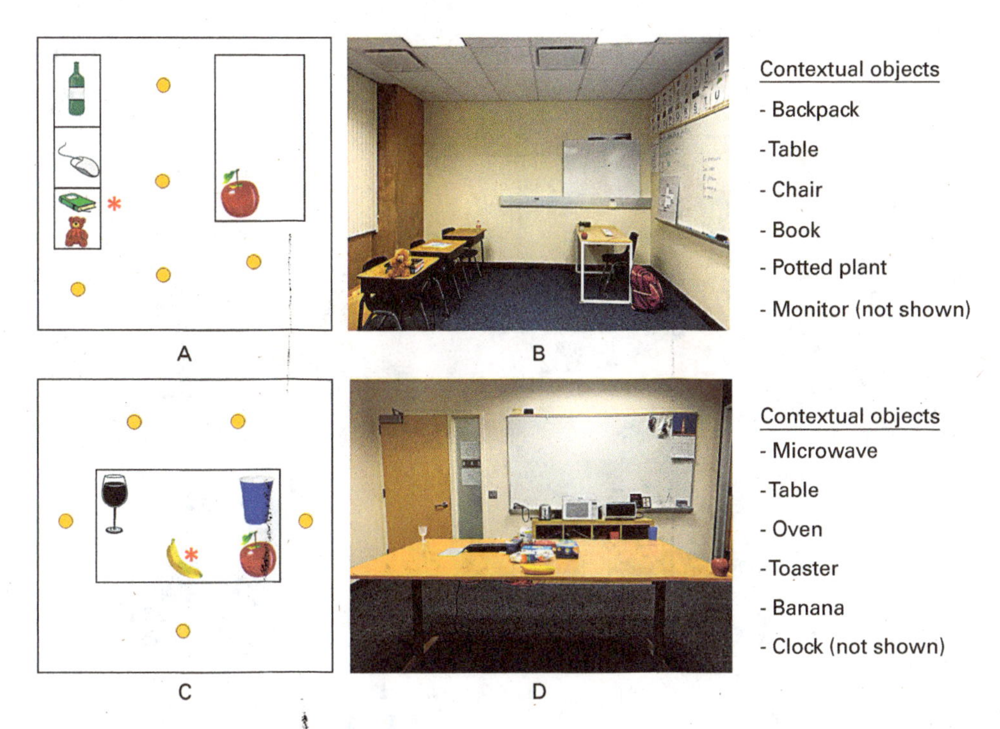

Figure 6.11
Experimental setup for the classroom and break room schemas. Yellow dots represent the destinations for the robot to roam and scan during training. *A*, Room layout and paired associations trained in the classroom. The bottle, teddy bear, and apple were laid out in the specified locations. After training on this layout, the teddy bear was exchanged for a computer mouse, introducing novelty. The robot then trained on this environment. The book (starred) was not trained as a paired association but was included as a contextual item during training. *B*, Physical setup of classroom and contextual items. *C*, Room layout and paired associations trained in the break room, which is adjacent to the classroom. The wine glass, cup, and apple were laid out on the central table. The banana (starred) was not trained as a paired association but was included as a contextual item during training. *D*, Physical setup of break room and contextual items.

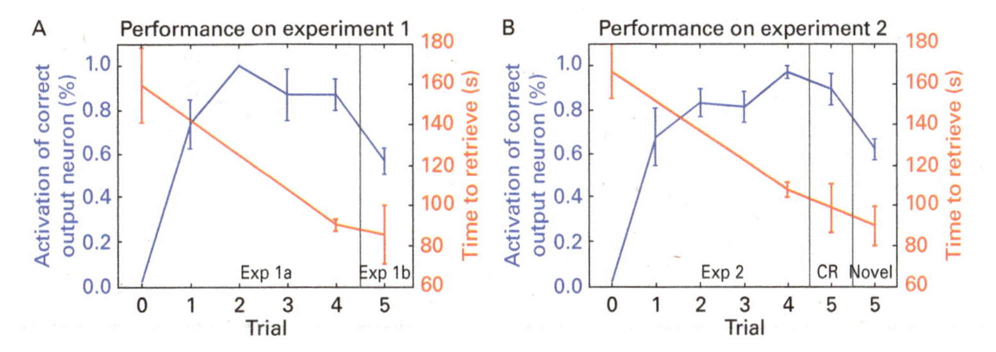

Figure 6.12
Performance on experiments 1 and 2 on a population of n = 5. *A*, For experiment 1, the activation of the correct location neuron of a cued object increased with training (blue line) and the retrieval time decreased (red line). When a novel object was introduced (Exp 1b), the location activation was still high and retrieval time was low, despite having had only one trial of training. *B*, Performance also improved over time when a novel schema was introduced in experiment 2. On returning to the classroom schema (CR), performance of original objects and novel objects was retained. Points denote the mean performance and error bars denote the standard deviation of the population.

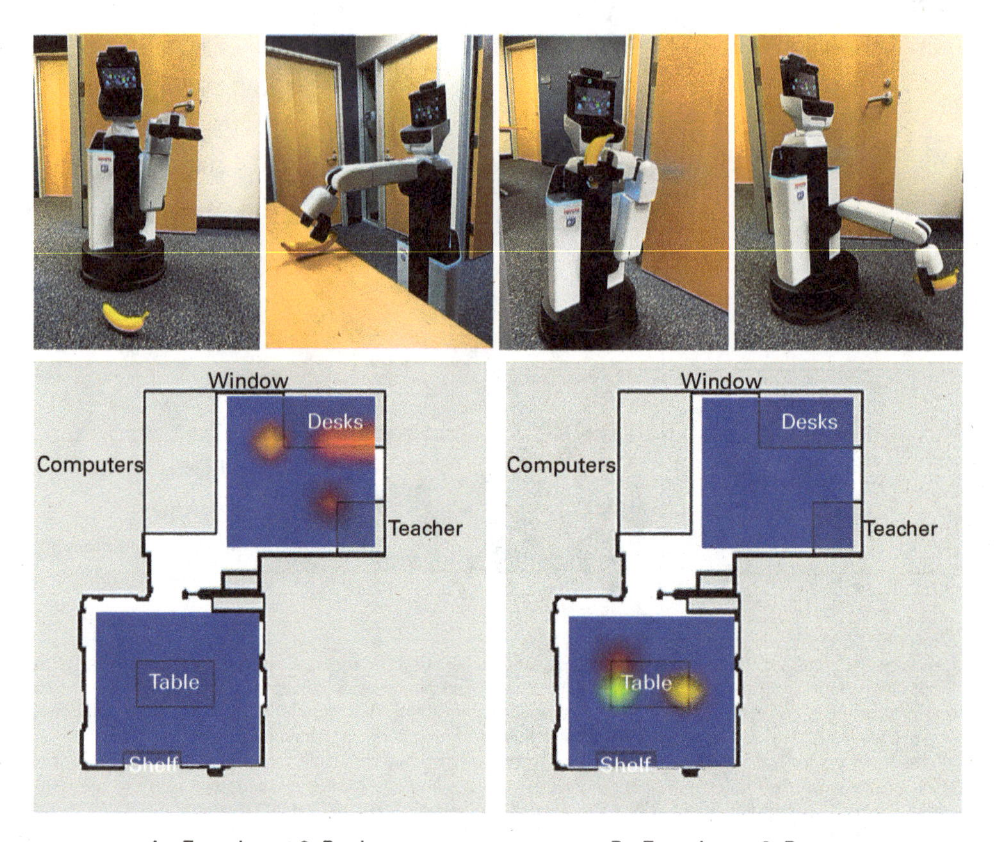

Figure 6.13
Cuing the robot on objects it had not retrieved before. *Top*: The robot was shown a banana. The robot then went to the break room to pick up the banana and navigated to the drop-off location to deposit it. *Bottom*: Heat map of action layer during experiment. *A*, When the robot was presented with a book and asked to retrieve it, the action layer showed high activity for objects in the classroom schema. *B*, When the robot was presented with a banana and asked to retrieve it, the action layer showed high activity for objects in the break room.

figure 6.13 show the HSR retrieving the banana during this experiment. The bottom of figure 6.13 shows the action layer activity in the neural network model indicating that the HSR understood which schema these objects belonged in and where they could be found.

These neurorobotic experiments show how context is tied to spatial representations via hippocampal interaction with the medial prefrontal cortex. Moreover it demonstrates how ideas from memory models in the brain may improve robotic applications and issues in artificial intelligence, such as catastrophic forgetting and lifelong learning.

The Toyota HSR is designed to aid humans in household tasks. In many tasks around the home, office, or classroom, context is important in determining how to carry out a task. Maintaining schemas of different rooms allows the robot to be aware of contexts. When asked to retrieve items at an unknown location, the robot explored the rooms associated with the appropriate schema. This led to faster retrieval times. The same idea could be applied to other tasks, such as tidying up, seeking individuals, or behaving appropriately given a context (e.g., being quiet in the classroom or raising one's voice in a noisy cafeteria).

6.6 Summary and Conclusions

Being able to adapt, retain information, and express knowledge are key elements for cognitive agents. For a neurorobot to claim itself as a cognitive agent, it needs these capabilities. In this chapter we discussed three key design principles that achieve these goals. First, the neurorobot must be able to learn and remember. In chapters 3 and 4 we described a number of learning rules that allow a neurorobot to learn from experience. The neurorobot examples in this chapter use variants of these learning rules to improve performance. Through their behavior they demonstrate that they remember what they have learned. Second, a form of learning related to reinforcement learning that is value-dependent is necessary to shape and adapt behavior. Value can be rewards or punishments received from the environment. It can also be internal, which has been called intrinsic motivation. Merely discovering something new might have value. Third, predicting outcomes is necessary for adaptive behavior and follows from the first two design principles. Predictions are based on past experiences, which implies that these experiences were learned and remembered. The memory is in the form of an internal model. Often what is to be predicted is value. When there is a mismatch because the expected value does not match the actual value, learning occurs to better fit the internal model to the current situation.

Because neurorobots are embedded in the real world, they can learn from experience, obtain extrinsic and intrinsic value through exploration, and build predictions to improve future performance. Built on the design principles in chapters 5 and 6, these neurorobots are a step closer to being truly cognitive machines. Rather than being purely reactionary, they demonstrate fluid, adaptable behavior that we commonly associate with many animals.

In the next chapter we discuss behavioral trade-offs that govern the survival of biological organisms and that can be used to endow neurorobots with complex, lifelike behavior.

7 Neurorobot Design Principles 3: Behavioral Trade-Offs Because Life Is Full of Compromises

7.1 Introduction

Biological organisms need to consider many trade-offs to survive (see table 7.1). These trade-offs regulate basic needs such as whether to forage for food, which might expose oneself to predators, or hide in one's home, which is safer but does not provide sustenance. These trade-offs can appear in cognitive functions such as introverted or extroverted behavior. Incredibly, many of these trade-offs are regulated by chemicals in our brain and body, such as neuromodulators or hormones. As we have discussed previously, these modulatory areas monitor and regulate environmental events. They send broad signals to the brain that can dramatically change behaviors, moods, decisions, etc. The brain can control these modulatory and hormonal systems by setting a context or making an adjustment when there are prediction errors.

In this chapter we explore several of these behavioral trade-offs. We discuss the neuroscience behind the trade-offs and neurorobots that incorporate these trade-offs. We consider these behavioral trade-offs to be neurorobotic design principles. By applying them to neurorobots, we may realize behavior that is more interesting and more realistic.

7.2 Reward versus Punishment

In previous chapters we described reinforcement learning rules and neurorobots that attempt to maximize rewards and minimize punishments. Dopaminergic neurons have responses that match quite well an error signal that is key for adjusting our current predictions about the world (Schultz et al., 1997). In particular, the phasic responses of dopamine neurons carry a **reward prediction error**, which is commonly used to update value expectations in reinforcement learning. What about punishment? One model suggested that tonic serotonin tracked the average punishment rate and that tonic dopamine tracked the average reward rate (Daw et al., 2002). They speculated that a phasic serotonin signal might report an ongoing prediction error for future punishment. It has been suggested that the serotonergic and dopaminergic systems activate in opposition for goal-directed actions (Boureau & Dayan, 2011).

Table 7.1
Hormones and neuromodulators that regulate behavioral trade-offs

Behavioral trade-off	Neural chemical	References
Reward vs. punishment	Dopamine, serotonin	Boureau and Dayan (2011); Cox and Krichmar (2009)
Invigorated vs. withdrawn Risky vs. conservative	Dopamine, serotonin	Boureau and Dayan (2011); Krichmar (2013)
Expected uncertainty vs. unexpected uncertainty	Acetylcholine, noradrenaline	Avery et al. (2012); Bouret and Sara (2005); Yu and Dayan (2005)
Explore vs. exploit	All neuromodulators	Aston-Jones and Cohen (2005); Hasselmo and McGaughy (2004); Krichmar (2008)
Feed vs. fight Forage vs. nest	Orexin	Aponte et al. (2011); Hahn et al. (1998); Luquet et al. (2005)
Stress vs. calm	Glucocorticoids	Rodrigues et al. (2009); Sapolsky (1996, 2015)
Social vs. solitary	Oxytocin, serotonin	Canamero et al. (2006); Dolen et al. (2013); Rilling and Young (2014); Tops et al. (2009)

This trade-off between reward and punishment can be quite nuanced when invigoration of activity can lead to rewards and punishment can lead to inaction (see figure 7.1).

Modeling has shown that direct anatomical opponency between these systems is not necessary to achieve behavioral opponency (Asher et al., 2010; Zaldivar et al., 2010). In many cases there is an environmental trade-off between the expected rewards and costs, and this can lead to interesting behavior. Asher and colleagues investigated this behavioral trade-off with a socioeconomic game known as hawk-dove. In the hawk-dove game players must choose between being aggressive by *escalating* a fight or peaceful by *displaying* a willingness to cooperate. Hawk-like aggression (*escalate*) can gain resources if the other player is peaceful (*display*). If both players escalate, there is a chance of being injured. If both players display, the resource is shared. In this way there is an expected reward and expected cost for these actions, which in the model was tracked by dopamine and serotonin, respectively. The neural network model shown in figure 7.2*A* controlled a computer agent's behavior. In simulations, this neural agent played against algorithmic agents that escalated with a set probability (*statistical*), a *tit-for-tat* strategy, or a *win-stay, lose-shift* strategy. The neural agent became aggressive playing against opponents when the chance of serious injury was low (top three charts in figure 7.2*B*). However, if the opponent was aggressive or the chance of serious injury was high, the neural agent was more cooperative (bottom three charts in figure 7.2*B*). The neural agent model was also deployed in human-robot interaction experiments (Asher et al., 2012). Interestingly, when this neurorobot was made more aggressive by artificially lesioning its serotonin system, human players tended to retaliate against the robot, which suggests that they thought the robot was not acting fairly.

Although this model is simple, it shows how interactions with the environment and other agents can lead to complex behavior. These and other robot and simulation studies

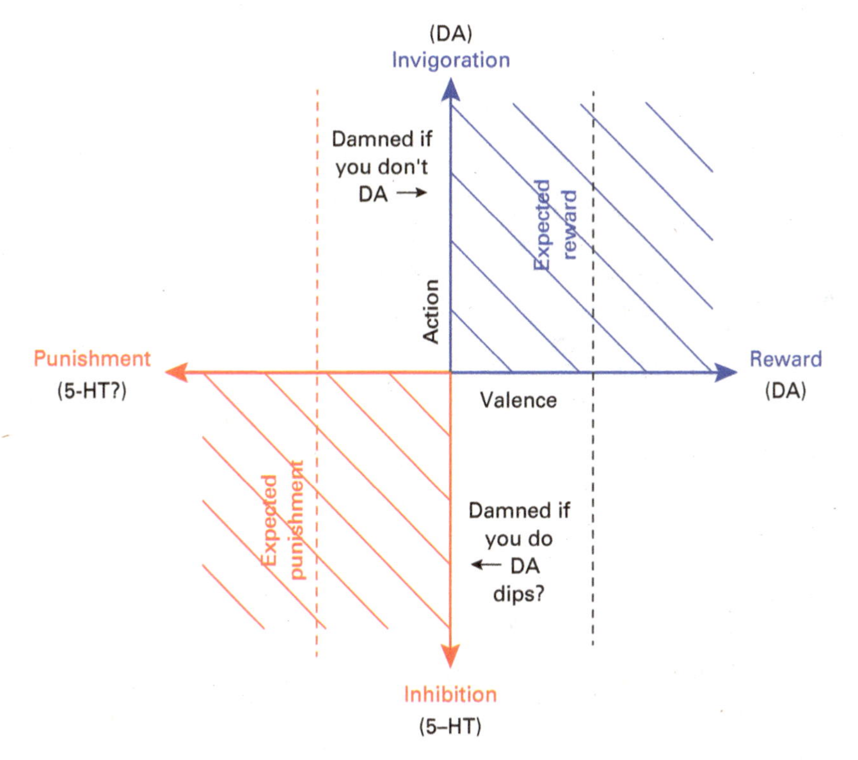

Figure 7.1
Behavioral trade-offs due to dopamine (DA) and serotonin (5-HT) opponency. The top-right quadrant is associated with better-than-expected outcomes (rewards) and leads to increased activity (invigoration). The bottom-left quadrant is associated with worse-than-expected outcomes (punishments), which can lead to decreased activity (inhibition). The other quadrants involve more complex interactions between reward and punishment. From Boureau and Dayan (2011).

suggest that interaction with the environment or with other agents can lead to interesting behavioral trade-offs. These trade-offs do not need to be explicitly modeled; they can emerge through behavioral responses in dynamic situations.

7.3 Invigorated versus Withdrawn

The dopamine and serotonin systems also regulate a trade-off between invigorated novelty seeking and withdrawn risk-averse behavior (see figure 7.1). It has been suggested that serotonin modulates the desire to withdraw from risk, which can take place in social interactions or in foraging for food (Tops et al., 2009). One can imagine that too much withdrawal from society could lead to symptoms of depression.

Consider the open-field test that is used to measure anxiety in rodents (Fonio et al., 2009). Usually, when a mouse is placed in an unfamiliar open area it will first stay near the borders

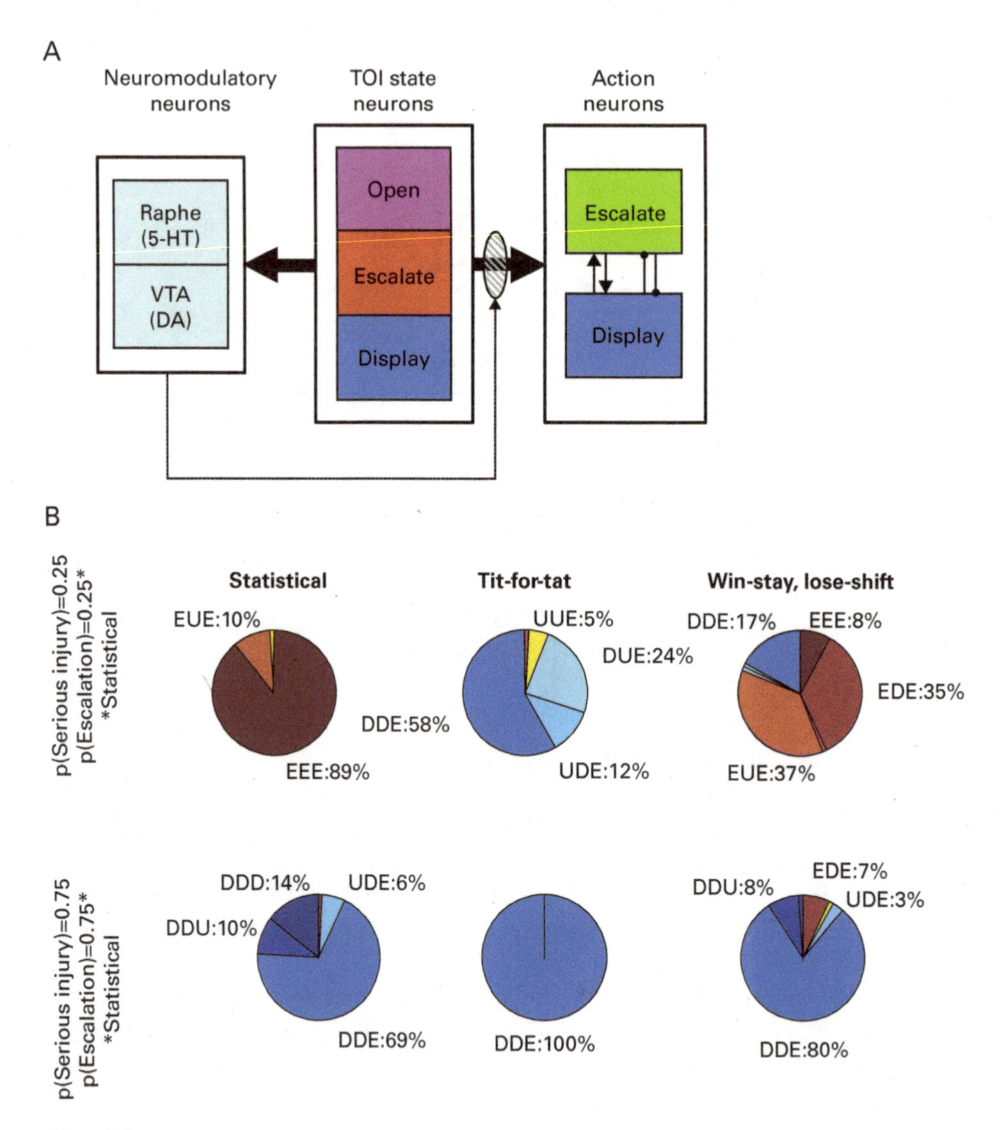

Figure 7.2
A, Agent's neural model to examine reward and punishment behavioral trade-offs. There were two neuromodulatory neurons, one representing serotonin (Raphe) and the other representing dopamine (VTA). There could be three target-of-interest (TOI) states taken by the other agent: open, escalate, and display. The neural model agent could take one of two actions, escalate or display. *B*, The pie charts show the proportion of probable actions taken by the neural agent in 100 games. The neural agent can commit to escalate (E), display (D), or undecided (U). Undecided represents a random choice between E and D. The labels represent the neural agent's response to the three TOI state areas. Dovelike strategies are displayed in blue, hawklike strategies in red, and arbitrary strategies in yellow. From Asher et al. (2010).

of the environment in which it might be concealed. The mouse may hide in a nest area if available. After some time the mouse decides the environment is safe and becomes curious. The mouse will then proceed to explore the environment by moving more and investigating the middle of the area. Serotonin levels can alter this behavior. For example, Heisler and colleagues showed that mice with increased serotonin spent less time in the center of the open-field arena (Heisler et al., 1998). In contrast, cocaine, which increases the level of dopamine, increased locomotive activity and the exploration of novel objects (Carey et al., 2008).

A neurorobot experiment took this trade-off into consideration by modeling the interactions between the serotonergic and dopaminergic systems (Krichmar, 2013). The algorithm was implemented in a neural network that controlled the behavior of an autonomous robot and tested in the open-field paradigm. When serotonin levels were high, sensory events led to withdrawn anxious behavior such as wall following and finding its nest (i.e., the robot's charging station). When dopamine levels were high, sensory events led to riskier reward-seeking behavior such as moving to the center of the arena or investigating novel objects. More details on this neurorobot experiment are discussed in the case study in section 7.9.

7.4 Expected Uncertainty versus Unexpected Uncertainty

The world is full of uncertainty with which we must cope in our daily lives. Sometimes the uncertainty is expected, forcing us to increase our concentration on a task. At other times the uncertainty is unexpected, forcing us to divert our attention. How we deal with these types of uncertainty can be thought of as a behavioral trade-off. Once again, neuromodulators influence this trade-off of how we apply our attention. Cholinergic neuromodulation is thought to track expected uncertainty (i.e., the known unreliability in the environment), and noradrenergic neuromodulation is thought to track unexpected uncertainty (i.e., observations that violate prior expectations) (Yu & Dayan, 2005). Yu and Dayan suggested that the basal forebrain, in which cholinergic neurons reside, encodes the uncertainty of prior information and that this can modulate attention to different features. They suggested that the locus coeruleus, in which noradrenergic neurons reside, is involved in cognitive shifts in response to novelty. When there are strong violations of expectations, locus coeruleus activity may induce a "network reset" that causes a reconfiguration of neuronal activity that clears the way to adapt to these changes (Bouret and Sara, 2005). In modeling and in experimental work, it has been shown that the cholinergic system mediates uncertainty seeking (Belkaid & Krichmar, 2020; Naude et al., 2016). Uncertainty seeking is especially advantageous in situations when reward sources are uncertain. The trade-off between expected and unexpected uncertainty can also be observed in how we apply our attention (Avery et al., 2012). Top-down attention or goal-driven attention, which ramps up our attention to look for something, is related to expected uncertainty. Bottom-up or stimulus-driven attention occurs when a surprise or unexpected uncertainty diverts our attention.

The Toyota HSR robot discussed in section 6.3.1 explored this trade-off between expected and unexpected uncertainty by modeling the cholinergic and noradrenergic system to regulate attention (Zou et al., 2020). Because the user's goals could be uncertain, simulated cholinergic neurons tracked how likely the user would be to choose any of these goals (i.e., expected uncertainty). When the user interacting with the robot changed their goals (i.e., unexpected uncertainty), the noradrenergic system in the model responded by resetting prior beliefs and rapidly adapting to the new goal. Figure 7.3 shows how the robot correctly guessed the user's goals, which then drove attention to the object associated with the goal (e.g., *eat* leads to attention to an apple or orange). Note how quickly the noradrenergic (NE) system responded to goal changes, which led to the cholinergic system (ACh) increasing attention to objects related to the new goal (see figure 7.3). We called this goal-driven perception, and in general the model was able to monitor a trade-off between the known and unknown uncertainties in the world to rapidly respond to the user's changing needs.

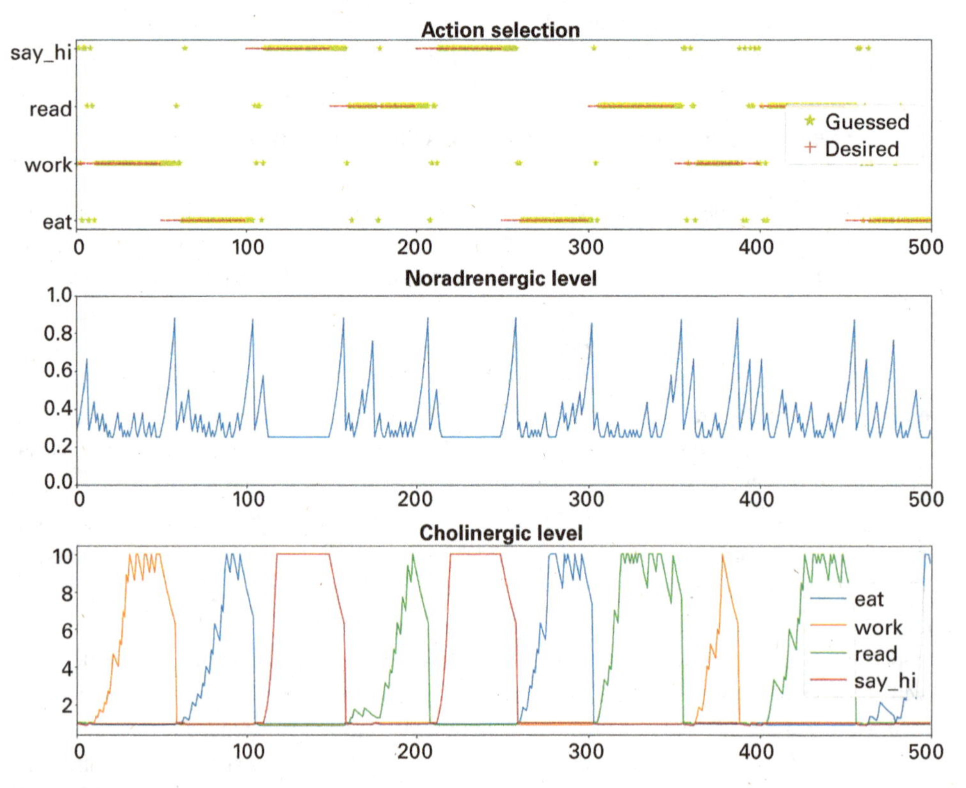

Figure 7.3
Goal-driven perception in a neurorobot. *Top chart*: Robot's response to guess the user's goal. *Center chart*: Noradrenergic (NE) neuron activity level. *Bottom chart*: Cholinergic (ACh) neuron activity level. There was an ACh neuron for each potential goal. From Zou et al. (2020).

7.5 Exploration versus Exploitation

During decision making or information gathering, there exists a trade-off between exploration and exploitation in which it is sometimes best to explore new options and other times it is best to exploit opportunities that have paid off in the past. A framework was presented in which neuromodulation controlled the exploration/exploitation trade-off (Aston-Jones & Cohen, 2005; Krichmar, 2008). When neuromodulators have tonic activity, the animal's behavior is exploratory and somewhat arbitrary. However, when the neuromodulator has a burst of phasic activity, the animal is decisive and exploits the best potential outcome at the given time. The CARL robot described in chapter 6 (see figure 6.5) incorporated tonic and phasic neuromodulation (Cox & Krichmar, 2009). When there was tonic neuromodulation, the robot randomly explored its environment by looking at different colored panels. If one of the colors became salient due to a reward or punishment signal, the robot's neuromodulatory systems responded with a phasic burst of activity. This phasic neuromodulation caused a rapid exploitive response to investigate the colored panel. It also triggered learning to approach positive-value objects and avoid negative-value objects.

7.6 Foraging versus Defending

Hormones are chemical messengers in the body that can affect the brain and other organs. They regulate a number of bodily functions such as body temperature, thirst, hunger, and sleep. Like neuromodulators, hormones can be triggered by environmental events and can broadly change neural activity. For example, the hormone orexin regulates hunger levels. This can lead to a behavioral trade-off in which animals foraging for food are less willing to defend a territory (Padilla et al., 2016). Foraging for food may cause an animal to leave its nest exposed to predators. However, defending one's territory requires energy expenditure, which if prolonged requires food for replenishment.

Hormones can be modeled and embodied in robots to explore interesting naturalistic trade-offs (Cañamero, 1997). For example, Cañamero's group modeled hormones that tracked a robot's health (Lones et al., 2018); one hormone was related to the battery level, and another hormone monitored the robot's internal temperature, which was related to how much the robot moved and the climate of its environment (see figures 7.4*A* and 7.4*B*). The robot's tasks were to maintain health and gather food resources, which might require aggressive action.

This neurorobotics study demonstrated how maintaining health requires behavioral trade-offs. Finding food increased the robot's internal temperature and reduced the robot's battery level. Being aggressive to obtain food also reduced battery levels. However, not searching for food would lead to starvation. Modeling the secretion and decay of hormones allowed the robot to maintain a comfortable energy and internal temperature level and at times led to an aggressive behavior of pushing objects away to get at food resources. However, as

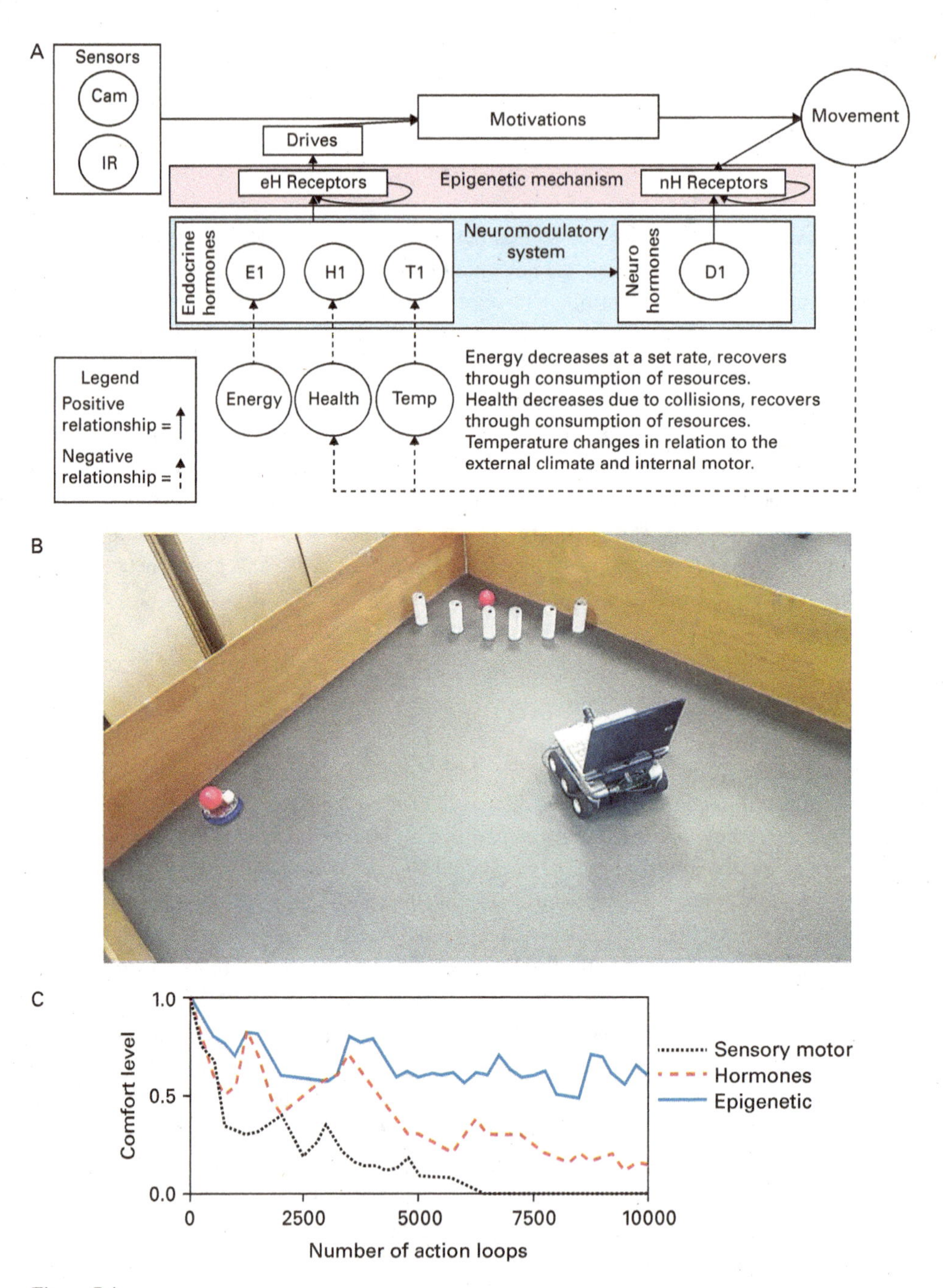

Figure 7.4
Hormone-driven, epigenetic robot. *A*, Robot architecture with its three interacting *systems*: basic architecture, neuromodulatory system, and epigenetic mechanism. *B*, Robot environment included energy resources in pink that could be moving (*left*) or stationary (*top right*). Objects that blocked access to resources could be pushed away. *C*, Performance in the most complex environment in which resources were available only for a limited time. From Lones and Canamero (2013).

the environments became more complex an epigenetic system, which monitored and controlled the hormones, became necessary for the robot's comfort level to be maintained satisfactorily (see figure 7.4*C*). Their epigenetic system acted in a similar way to the **hypothalamus**, a subcortical brain region that regulates many of our bodily functions. Their experiments showed that an epigenetic mechanism significantly and consistently improved the robot's adaptability and might provide a useful general mechanism for adaptation in autonomous robots.

7.7 Stress versus Calm

In his book *Why Zebras Don't Get Ulcers*, Robert Sapolsky describes how a zebra, which is calmly grazing, responds when it encounters a lion (Sapolsky, 2015). The zebra quickly runs away from this stressful situation. Once clear of danger, the zebra is calm again. This fight-or-flight response is mediated by the stress hormones known as **glucocorticoids**, which increase blood flow and awareness. However, this stress response does come at the expense of regulating long-term health and short-term memory (Chiba & Krichmar, 2020). Unlike zebras, people sometimes remain in a constant state of stress due to elevated glucocorticoids, which can cause damage to the hippocampus and memory.

Although there has been little work to date on neurorobots that regulate their stress level, downregulation of behavior could be useful for autonomous systems far from power sources, which might have a stress-like response to carry out a mission and then switch to a low-power calm mode after the mission has been accomplished.

In an interesting paradigm that explores the stress versus calm trade-off, experiments have shown that rats appear to be capable of empathy and prosocial behavior. In one study a rat was trapped in a cage and clearly stressed (Ben-Ami Bartal et al., 2011). Another rat observing this behavior became stressed, too. The observing rat, feeling bad for its trapped friend, found a lever that opened the cage and released the trapped rat. This study suggested that rats can feel another's pain (i.e., feel empathy) and are willing to act on the other's behalf (i.e., can be prosocial).

In a robotic variation of the empathy experiment, a rat trapped in a cage interacted with two different robots, one of which was helpful and opened the cage (figure 7.5*A–B*) and the other of which was uncooperative and ignored the trapped rat (Quinn et al., 2018). Interestingly, the rat remembered who its robot friends were. When the helpful robot was trapped, the rat freed that robot (see figure 7.5*C–F*) but did not free the robot that was uncooperative. This could have implications for assistive robotics. A robot that can identify and relieve stress or anxiety could have applications for robotic caretakers or for disaster relief.

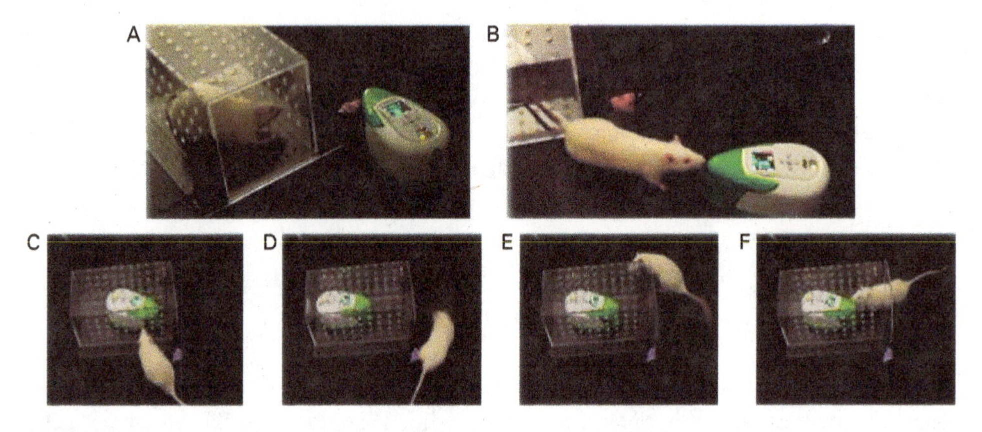

Figure 7.5
Empathy and relieving stress in a rat-robot interaction experiment. *A*, The helpful robot pushes the lever to release a rat from the restrainer. *B*, A typical nose-to-nose check from the rat following the robot's opening the door. *C*, A rat viewing the helpful robot trapped in the restrainer. *D*, The rat moves directly from viewing the trapped robot to the side of the restrainer where the lever is located. *E*, The rat presses the lever to open the door for the helpful robot (note the gaze of the rat toward the robot). *F*, The rat interacts with the robot immediately following door opening. From Quinn et al. (2018).

7.8 Social versus Solitary

Hormones can regulate a trade-off between social bonding and independence. Estrogen, progesterone, oxytocin, and prolactin can influence a number of neural systems to ensure maternal nurturing, bonding, and protection of young (Rilling & Young, 2014). In particular, **oxytocin** has been shown to regulate social and paternal bonding (Young & Wang, 2004). An interesting example of oxytocin's effect on bonding has been observed in voles. Whereas prairie voles are polygamous and the male does not assist in the nurturing of young pups, meadow voles are monogamous and both parents participate in the pup rearing. Interestingly, meadow voles have more oxytocin receptors than prairie voles. Furthermore, inhibiting oxytocin prevents pair bonding in meadow voles (Young & Wang, 2004). However, social bonding requires devoting energy to another, possibly at the expense of one's own health. Therefore it could be argued that there should be a balance between social and solitary behavior.

Neurorobot experimenters have investigated the balance between social and solitary behavior. By simulating social hormones, Cañamero's group investigated attachment bonds between a robot and a "caregiver" (Canamero et al., 2006; Hiolle et al., 2009; Hiolle et al., 2012). Although they did not explicitly model oxytocin, their system did simulate the type of bonding observed between a parent and a child (see figure 7.6). The robot found a good balance between asking the caregiver for help and learning on its own. Too much interaction with a caregiver led to stress and rejection by the robot. Not enough interaction with a caregiver resulted in isolation. As with humans, a proper balance is important for learning and development.

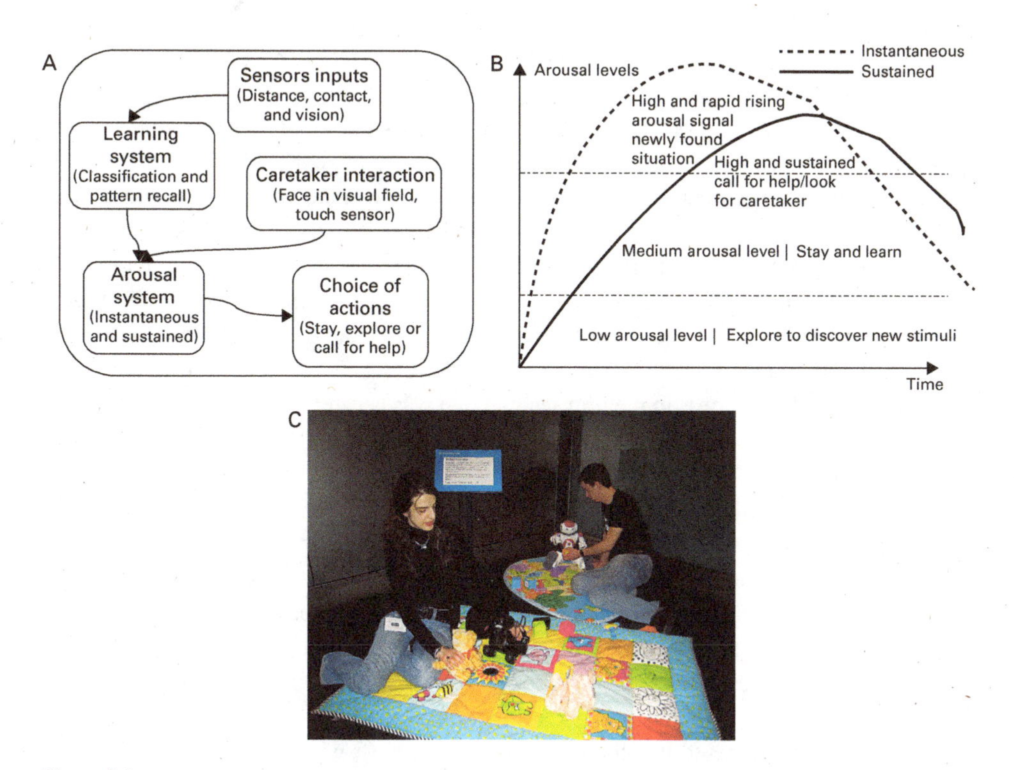

Figure 7.6
Caregiving behavior with human-robot bonds. *A*, Neural architecture for robot's arousal system. *B*, Arousal levels as the robot discovers novel items or works with a caretaker. *C*, Experimental setup with caretakers interacting with the Aibo dog and Nao humanoid. Adapted from Hiolle et al. (2009) and Hiolle et al. (2012).

7.9 Case Study: Anxious and Curious Behavior in a Neurorobot

We discussed previously that there is a behavioral trade-off between invigorated and inhibited activity and that this trade-off may be governed by dopamine (DA) and serotonin (5-HT), respectively (figure 7.1). Furthermore, we discussed how acetylcholine and noradrenaline can direct attentional resources. It is also known that the frontal cortex can influence cognitive control by gating on and off behaviors as well as by influencing neuromodulatory levels. In particular the orbitofrontal cortex (OFC) has strong interactions with the dopaminergic system (Frank & Claus, 2006), and the medial prefrontal cortex (mPFC) has strong interactions with the serotonergic system (Jasinska et al., 2012). This case study describes how putting these ideas together can lead to interesting, naturalistic behavior in a robot.

The interaction between neuromodulators and the frontal cortex were implemented in a neural network that controlled the behavior of an iRobot Create (Krichmar, 2013). The neural network ran on a small laptop that was mounted on the robot. The robot's touch sensors were used to detect bumps, the laptop's webcam to monitor light levels, and a laser range finder to detect objects.

Robot control was achieved through processing events and states. States were behavioral primitives, and events were driven by sensory signals. An event could cause a switching of behavior states. The neural simulation arbitrated between incoming events and decided when to switch states. A simulation cycle occurred approximately once per second, which was roughly the time needed to read the robot's sensors, update the neural simulation, and send a motor command. The neural network handled three sensory events: (1) Object detected. This event was triggered when its laser range finder detected an object between 12° and 30° wide and closer than 1 m. (2) Light detected. This event was triggered when the average pixel brightness in the grayscale image was greater than 50 percent. (3) Bump detected. This event was triggered by the robot's bump sensors or by the laser detecting an object closer than 20 cm.

The robot switched between four behavior states: (1) Wall follow (see figure 7.7*B*). This caused the robot to follow the wall to its right. (2) Find home (see figure 7.7*C*). This caused the robot to move in a random pattern until it detected the docking station via an IR beam that had a range of roughly 500 cm. (3) Open field (see figure 7.7*D*). The robot drove toward the most open area of the environment, as judged by the laser range finder. (4) Explore object (see figure 7.7*E*). The robot moved toward the object detected by the laser.

Sensory events triggered the neuromodulatory neurons, which in turn drove behavior states (see figure 7.8). Frontal cortex areas (see OFC and mPFC in figure 7.8) triggered action selection and exerted cognitive control on the neuromodulatory areas (see DA and

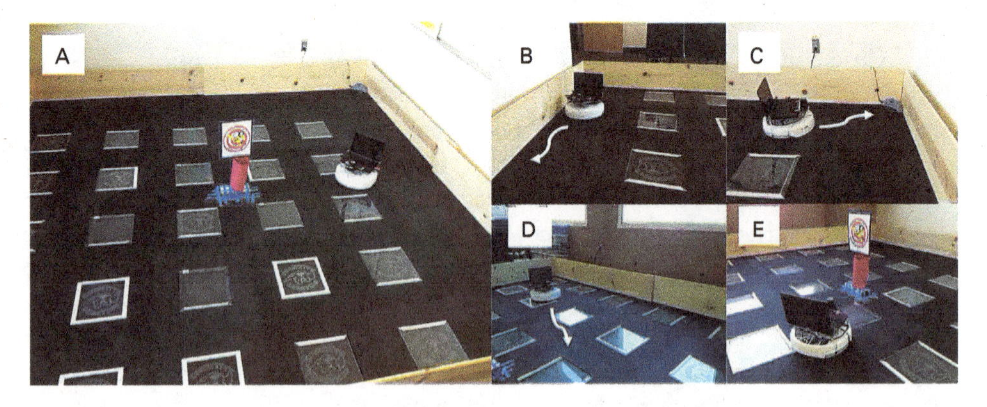

Figure 7.7
Setup for neurorobotic experiments. Experiments were run on an iRobot Create equipped with a URG-04-LX laser range finder (Hokuyo Automatic) and a System 76 netbook running the Ubuntu Linux operating system for computation. *A*, Environment was a 3.7 m^2 arena enclosed by plywood. The picture in the middle was a novel object for the robot to explore. *B*, Wall follow behavior. *C*, Find home behavior. Finding the robot's docking station. *D*, Open field behavior. The robot moved toward open spaces in the environment on the basis of laser range finder readings. *E*, Explore object. The robot approached narrow objects on the basis of laser range finder readings.

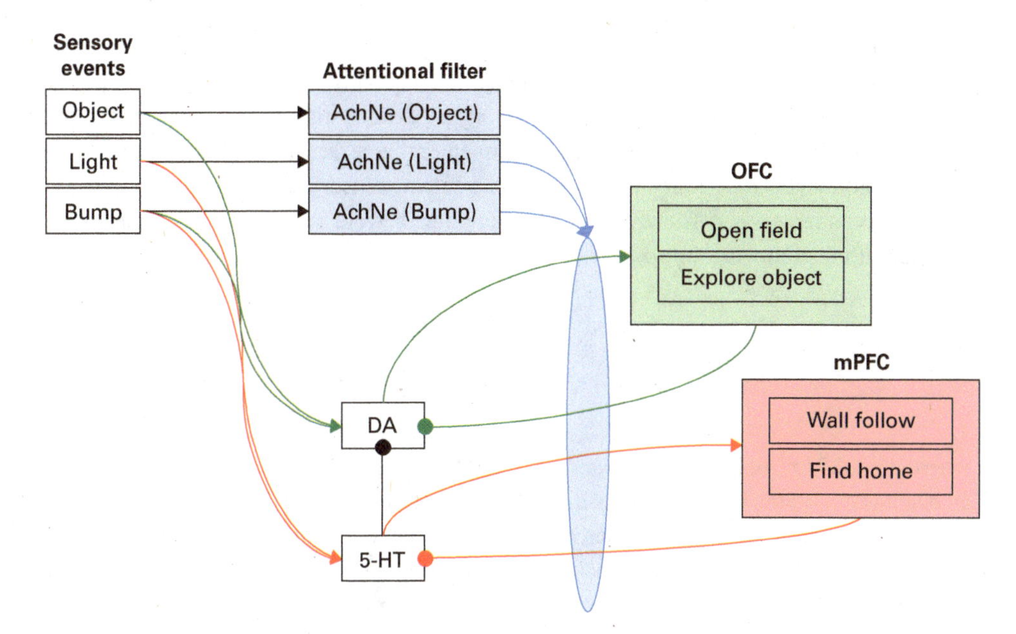

Figure 7.8
Neural architecture to control robot behavior. Sensory events were handled by three binary neurons. These neurons projected to the attentional filter neurons (AchNE) and the dopaminergic and serotonergic neurons (DA and 5-HT). The DA and 5-HT neurons projected to the OFC and mPFC neurons. The most active OFC or mPFC neuron dictated the robot's behavioral state. The AChNE neurons had a modulatory effect on the projection from the DA and 5-HT to OFC and mPFC (see blue ellipse and arrows). OFC and mPFC projected to 5-HT and DA neurons with inhibitory connections. Excitatory and inhibitory connections within and between OFC and mPFC neurons were all-to-all.

5-HT in figure 7.8) via inhibitory projections. The cholinergic (ACh) and noradrenergic (NE) systems (see AChNE in figure 7.8) acted as an attentional filter allowing novel and unexpected events to gate through to the frontal cortex. Specifically, AChNE modulates connections from DA and 5-HT to cortical neurons and inhibitory connections between cortical neurons (see blue arrows and ellipses in figure 7.8).

The neural simulation consisted of three event neurons, each of which corresponded to one of the sensory events described previously; four state neurons, each of which corresponded to one of the behavioral states described previously; and neuromodulatory neurons. There were one DA neuron, one 5-HT, and three AChNE neurons, each of which corresponded to one of the sensory events described previously. Figure 7.8 shows the architecture and connectivity of the network. Detecting an object with the laser signaled novelty or something potentially rewarding in the environment that was worth taking a risk to investigate. Therefore these events triggered dopaminergic neurons (Object → DA in figure 7.8). A bright light signaled a potential danger and thus triggered serotonergic neurons (Light → 5-HT in figure 7.8). A bump could signal something in the environment that was either interesting or noxious. Therefore, the bump event triggered both dopaminergic and serotonergic neurons (Bump → DA and Bump → 5-HT in figure 7.8).

AChNE neurons acted as an attentional filter for events by adjusting weights from event neurons to AChNE. AChNE neurons habituated to frequently occurring events by implementing a depressive short-term plasticity using an exponential decay. Tonic activity in the DA and 5-HT neurons was modeled by implementing a facilitative short-term plasticity. The tonic levels rose every time there was a salient sensory event and decayed exponentially between events. These rising and falling levels were set by an exponential decay function of the following form:

$$nm(t) = \begin{cases} p * nm(t-1); \text{ if an event occurs} \\ nm(t-1) + \dfrac{1 - nm(t-1)}{\tau}; \text{ otherwise,} \end{cases} \tag{7.1}$$

where p is 0.25 for AChNE neurons and 1.25 for DA or 5-HT neurons. For AChNE neurons an event is any of the sensory events described above. For DA and 5-HT neurons an event is gated by when AChNE neuron activity is greater than 0.5. Having p less than 1 results in a decrease every time an event occurs. For example, the first time the robot bumps into something there is a strong response. Every subsequent time, the bump has less effect on ACh/NE activity. Having p greater than 1 has the opposite effect. Every time an event is gated in, the tonic neuromodulatory response increases.

Event neurons were binary and set to 1 when an event occurred and 0 otherwise. All other neurons were governed by a sigmoid activation function, which kept neural activity between 0 and 1, as follows:

$$\frac{1}{1 + e^{-gI(t)}}, \tag{7.2}$$

where g was the gain of the function and I was the input to the neuron. More details on the model can be found in Krichmar (2013).

The experimental setup was designed to mimic a rodent open-field experiment. When placed in a new environment rodents typically stay near their nest (i.e., the docking station) or follow closely along the walls of an environment (Fonio et al., 2009). As they become more comfortable in the environment they will venture out into the open area of the arena or explore a novel object placed in the arena. This paradigm is often used to test animal models of anxiety (Heisler et al., 1998). The present experiments were designed to test how dopaminergic and serotonergic neuromodulation influence the ability to cope with a stressful event. Each experimental trial lasted 2 minutes. Experiments were conducted in the dark. Halfway through the trial the lights were flashed on and then off. A light flash was used to mimic a stressful event in the open-field test because rodents generally prefer the dark.

The robot responded appropriately to sensory events in its environment. Novel objects resulted in its exploring the environment, and stressful events caused the robot to seek safety. Figure 7.9*A* shows a representative trial in which there were balanced tonic levels of neuromodulation ($\tau_{DA} = \tau_{5HT} = 50$). The *x*-axis denotes time in seconds from the start of the trial until the end, which was approximately 240 seconds. The upper chart shows the robot's behavioral state over the course of the trial. The second through fifth charts show the neural activity of the state, event, AChNE, and neuromodulatory neurons, respectively, over the course of a trial where dark blue signifies no activity and bright red signifies maximal activity. The bottom chart shows the level of tonic neuromodulation. Note that initially when the environment was unfamiliar, 5-HT activity dominated, resulting in anxious behavior such as WallFollow and FindHome actions. However, once the robot had become more familiar with its environment (approximately 60 seconds into the trial) DA levels were higher and there was more curious or exploratory behavior. Note that the AChNE neurons gated only through interesting and rare events. This was achieved through AChNE modulation of projections from neuromodulatory neurons to OFC and mPFC and through AChNE modulation of intrinsic inhibitory projections between frontal cortex neurons. For example, constant bump events were habituated (compare bump event neuron activity with bump AChNE activity shown in figure 7.9*A*).

At approximately 120 seconds into the trial, there was an unexpected light event due to flashing the lights on and off, which resulted in a phasic 5-HT response and a longer tonic increase in 5-HT. This caused the robot to respond with withdrawn or anxious behavior until approximately 210 seconds into the trial when a pair of object events triggered exploration of the center of the environment. Specifically, tonic levels of 5-HT had decayed, and the object events caused an increase in DA levels triggering a change in behavioral state.

Figure 7.9*B* shows the proportion of curious behavior (OpenField and ExploreObject) and anxious behavior (FindHome and WallFollow) for five experimental trials. Each bar was the average proportion of time spent in either curious (green bars) or anxious (red bars) behaviors. The error bar denoted the standard error. Figure 7.9*C* shows the behavior time-locked to the light event. The light event, which occurred at approximately the halfway point in the trial, was introduced to cause a stress response. After the light event, the neurorobot's behavior rapidly switched to anxious behavior for roughly 60 seconds when it became curious again. Variation occurred due to different times of the light event and random variations in other sensory events.

It has been suggested that degradation of serotonin re-uptake can have detrimental effects on the ability to cope with stressors (Jasinska et al., 2012). To mechanistically test this notion, the time constant for tonic serotonin was increased ($\tau_{DA} = 50$, and $\tau_{5HT} = 150$). This had the effect of serotonin staying in the system longer after a stressful event. A stressful event such as a bright light still caused the robot to select anxious behaviors, but the increase in serotonin levels resulted in never breaking out of this stressful behavior.

Increasing the levels of DA by adjusting the tonic time constant ($\tau_{DA} = 150$, and $\tau_{5HT} = 50$) resulted in more curiosity and risk taking but did not abolish the stress response. For

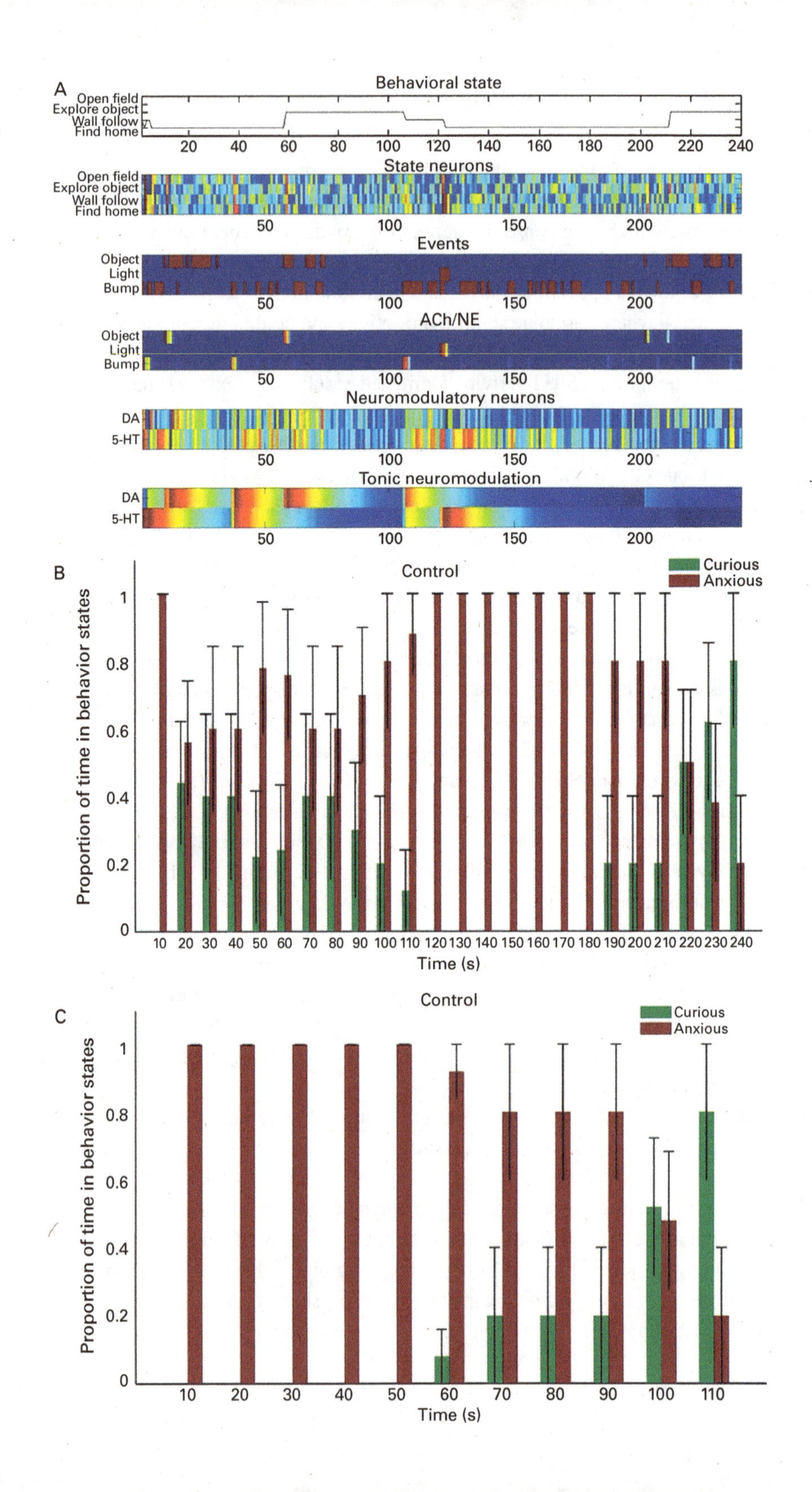
A
Behavioral state
Open field
Explore object
Wall follow
Find home
20 40 60 80 100 120 140 160 180 200 220 240
State neurons
50 100 150 200
Events
Object
Light
Bump
ACh/NE
Neuromodulatory neurons
DA
5-HT
Tonic neuromodulation
B
Control
Curious
Anxious
Proportion of time in behavior states
Time (s)
C
Control
Curious
Anxious
Proportion of time in behavior states
Time (s)

example, the light event did cause a strong increase in 5-HT activity, which in turn inhibited DA activity. However, the next sensory events, which were gated through by the AChNE attentional filter, resulted in strong DA activation and curiosity-seeking behavior. The population data reflected this interplay between the DA and 5-HT system. The robot responded to the stressful event but was much more curious than the controls. In effect, it took more risks by venturing into the middle of the environment during or right after the stressful light event. Similarly, cocaine, which increases levels of DA in the nervous system, has been shown to increase activity in the open-field test with rats as well as increase the exploration of novel objects (Carey et al., 2008).

The OFC and mPFC areas of the model exerted cognitive control on behavior by inhibiting the DA and 5-HT systems, respectively. Activity in OFC and mPFC initiated behavior selection but also inhibited the neuromodulatory systems. This inhibition kept the appropriate neuromodulatory system in check and exerted cognitive control by signaling to the neuromodulatory system that the sensory event had been handled. When the projections from mPFC to 5-HT were lesioned in the model, the serotonergic system was overactive and the robot acted anxious almost entirely. In contrast, when the OFC to DA projection was lesioned, the robot was overly curious to the point that its behavior resembled obsessive-compulsive behavior.

The neurorobotic experiments presented in this case study demonstrate that the opposition of the serotoninergic system with the dopaminergic system can lead to the type of anxious and curious behavioral trade-off observed in animals. Whereas high levels of 5-HT led to withdrawn, anxious behavior by suppressing DA action, high levels of DA or low levels of 5-HT led to curious, exploratory behavior. Moreover, it was shown that top-down signals from the frontal cortex to these neuromodulatory areas were critical for handling both stressful and positive valence events. The action of the neuromodulatory system and its interaction with areas important for action selection and planning are in a fine balance. It was shown that if any of these systems become out of balance, due to lesions or changes to the efficiency of neuromodulatory signaling, aberrant behavior occurs. This may have implications for understanding mood disorders, obsessive-compulsive disorders, and anxiety.

Figure 7.9
Behavioral and neural responses in the intact model. *A*, Behavioral and neural responses in a representative trial. The *x*-axis for all charts shows the time of the trial in seconds. The chart labeled *Behavioral state* denotes the state of the robot at a given time. The charts labeled *State neurons*, *Events*, *ACh/NE*, and *Neuromodulatory neurons* show the neural activity over the trial, where dark blue equates to no activity and bright red equates to maximal activity. Note that event neurons were binary. The chart labeled *Tonic neuromodulation* denotes the level of tonic activation contributing to DA and 5-HT neurons. *B*, The proportion of curious (ExploreObject and OpenField) and anxious (FindHome and WallFollow) behavior averaged over five trials. The error bars denote the standard error. The histogram binned the behavior in 10 s windows. *C*, Similar to *B* except that the behaviors were time-locked to the light event.

7.10 Summary and Conclusions

The world is full of contradictions and variation that require us to make choices. Our changing needs lead to a number of trade-offs to assess. In this chapter we have discussed behavioral trade-offs such as (1) reward vs. punishment, (2) invigorated vs. withdrawn, (3) expected vs. unexpected uncertainty, (4) exploration vs. exploitation, (5) foraging for food vs. defending one's territory, (6) stress vs. calm, and (7) social vs. solitary.

One recurring theme for these trade-offs is that they are regulated at a very basic level in the nervous system. Subcortical sources of neuromodulators and hormones can strongly affect which side of the trade-off to choose. As has been discussed, the central nervous system monitors and influences the activity of these subcortical areas on the basis of top-down predictions or goals. Still, it is interesting that many behaviors that we associate with cognition and free will are the result of chemical concentrations.

Another recurring theme in this chapter is how modeling trade-offs can lead to interesting behavior. As any storyteller would tell you, conflict makes for a good story. There is a reason why the characters in bestsellers and blockbuster movies are often conflicted. Their conflicts are what make the story interesting and are also to a lesser degree experienced in our own lives. Implementing some of these behavioral trade-offs in our robots can make them more interesting and relatable.

III NEUROROBOT APPLICATIONS

In this part we provide more examples of neurorobots in action. Navigation is an important area of research in robotics and in neuroscience. Therefore we provide a chapter on neurorobotic approaches to navigation. Cognitive science and neuroscience strongly influence the fields of developmental robotics and social robots. We discuss examples of these types of robots and ground them in what is currently known from neuroscience. Finally, the book wraps up with a summary and speculation on where we see neurorobotics heading in the future.

8 Neurorobotic Navigation

8.1 Introduction

Although navigation is not necessary for survival, it increases the chances of it. Compare, for instance, living agents that are unable to move with agents that can move and plan their movements. Agents that are unable to move are at the mercy of the environment, passively gathering whatever resources are available. Agents that move without planning can react to short-term events, reflexively traveling towards resources and away from dangers. Agents that can plan their movements across space and time are able to expand their access to a far wider range. Spatial navigation consists of two basic components: a map of space and a means of using that map to plan future movements. The implementations and strategies for these components differ between traditional robotics and neurobiology. However, they both are bound by physical limitations and energy needs. In this chapter, we cover mapping and planning in neurorobotics, the influences of traditional robotics, the neurobiological details of spatial representation and planning, and the merging of the two in neurorobotic navigation.

8.2 Mapping

A map of the environment contains different features, depending on the organism's goals and how the organism travels. For example, a flying animal or robot would want a 3D representation of space and would worry less about ground obstacles, whereas a wheeled robot operating indoors would be highly aware of obstacles and passable areas such as walls and doors. Objects in a map can be further split into static and dynamic categories. Static objects are permanent obstacles that do not change much and can be stored in memory for long-term use, and dynamic objects are temporary and require constant online sensing to detect. Equipped with a map, an agent must be able to sense its surroundings and figure out where it is located on the map. This requires the integration of sensory information and the agent's history, which accounts for where it has been recently to determine where it is now. A further challenge for the agent is to learn and build a new map of a previously unseen area. There have been many interesting approaches, artificial and biological, that have addressed these challenges in mapping.

8.2.1 Mapping Challenges in Traditional Robotics

A longstanding challenge in traditional robotics is how to learn a map of a new environment, localize oneself within the map, and do both tasks simultaneously. An important robotics research area is dedicated to developing **simultaneous localization and mapping** (SLAM) algorithms. In this chapter we cover some details of SLAM and other traditional robot mapping and path planning algorithms. For a more in-depth treatment, we refer the interested reader to the textbook by Choset et al. (2005).

A common technique for mapping is the maintenance of an **occupancy grid**. To create an occupancy grid, an area is represented as a grid of squares initialized to zero values. The center of the grid can be initialized as the robot's current position. If the robot is equipped with a distance sensor such as a sonar or light detection and ranging (lidar), the sensor can perform measurements within the robot's field of view, finding the distances of the closest obstructions. By using the distance measurements and angular positions with respect to the robot's position, the occupancy grid can be formed by populating the corresponding grid locations. Figure 8.1 shows an example of this. After the robot performs several of these measurements in various parts of the area, the occupancy grid eventually represents a map of the area. It is important to have full exploration of the area. As an example, if the robot explores only one side of a wall, the entire unexplored portion of the map will be unknown to the robot. Moreover, multiple scans of the environment can help determine if there are any dynamic objects in an area. If the robot detects an obstacle at one point in time but does not detect it the next time, it can be considered a dynamic object and not included in the map. This mapping technique assumes that the robot's controls and state measurements are perfect, that is, if the robot is commanded to move forward and turn by a certain amount, this can be performed without any error or slippage. In practice, this is rarely the case.

Given a learned map of the environment, localizing oneself within the map is another challenge. **Markov localization** is one technique for localization. Figure 8.2 shows an example of Markov localization. The map consists of a long hallway with three doors, unevenly spaced. The robot has a belief probability distribution about its own position x, represented as $bel(x)$. As the robot moves and senses, it continually updates its belief distribution with new information. For instance, when the robot starts in a new environment, it has no belief about its location because it has not moved or sensed anything. After the robot senses the new door, it incorporates a new probability, $p(z|x)$, in which z is the event of sensing a door when given the robot's location x. Because the probability of sensing a door is high at the locations of actual doors, the robot's belief becomes a bump distributed around each of the door locations equally. The robot then moves forward down the hallway. The belief of the robot's position then moves forward, represented as a bump ahead of each door but with a lower height than before due to the increased uncertainty. The robot

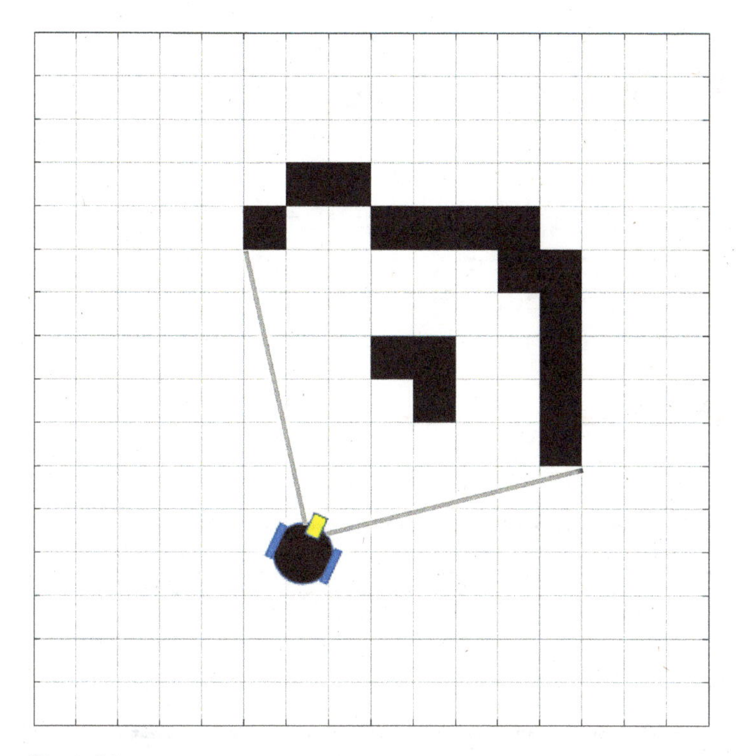

Figure 8.1
Example of occupancy grid construction. The robot, shown in the lower left, uses a distance sensor to determine which parts of a 2D grid space are occupied by an impassable area such as a wall or obstacle. The gray lines emanating from the robot denote the area the robot's distance sensor sweeps. The impassable areas of the grid are shown in black.

then senses the second door. Combining the belief of being near a door with its previous belief, the robot then has a very strong belief about its location next to the second door. The same principle can be applied to a larger 2D or 3D environment.

SLAM is the challenge of being able to construct a map of a new environment while simultaneously locating oneself in the map. Pure mapping tasks assume no error from the robot's motion, and pure localization tasks assume no error from the map. However, neither case is realistic, especially when the robot is in a new environment. Any mechanical robot in a physical space is bound to have errors; that is, when it plans to rotate or move forward by a certain amount, the actual ending position will not be exactly as expected.

Dead reckoning is the act of navigating from one's recently known positions using one's own actions. For example, if one is sailing in open ocean on a cloudy day, the boat's position can be estimated using the compass heading and the boat's velocity. Dead reckoning relies heavily on the accuracy of the self-sensing apparatus. In robotics this might be odometry or an inertial measurement unit (IMU), which uses an accelerometer, gyroscope, and compass, to measure the robot's self-movement. While highly accurate robots may be able

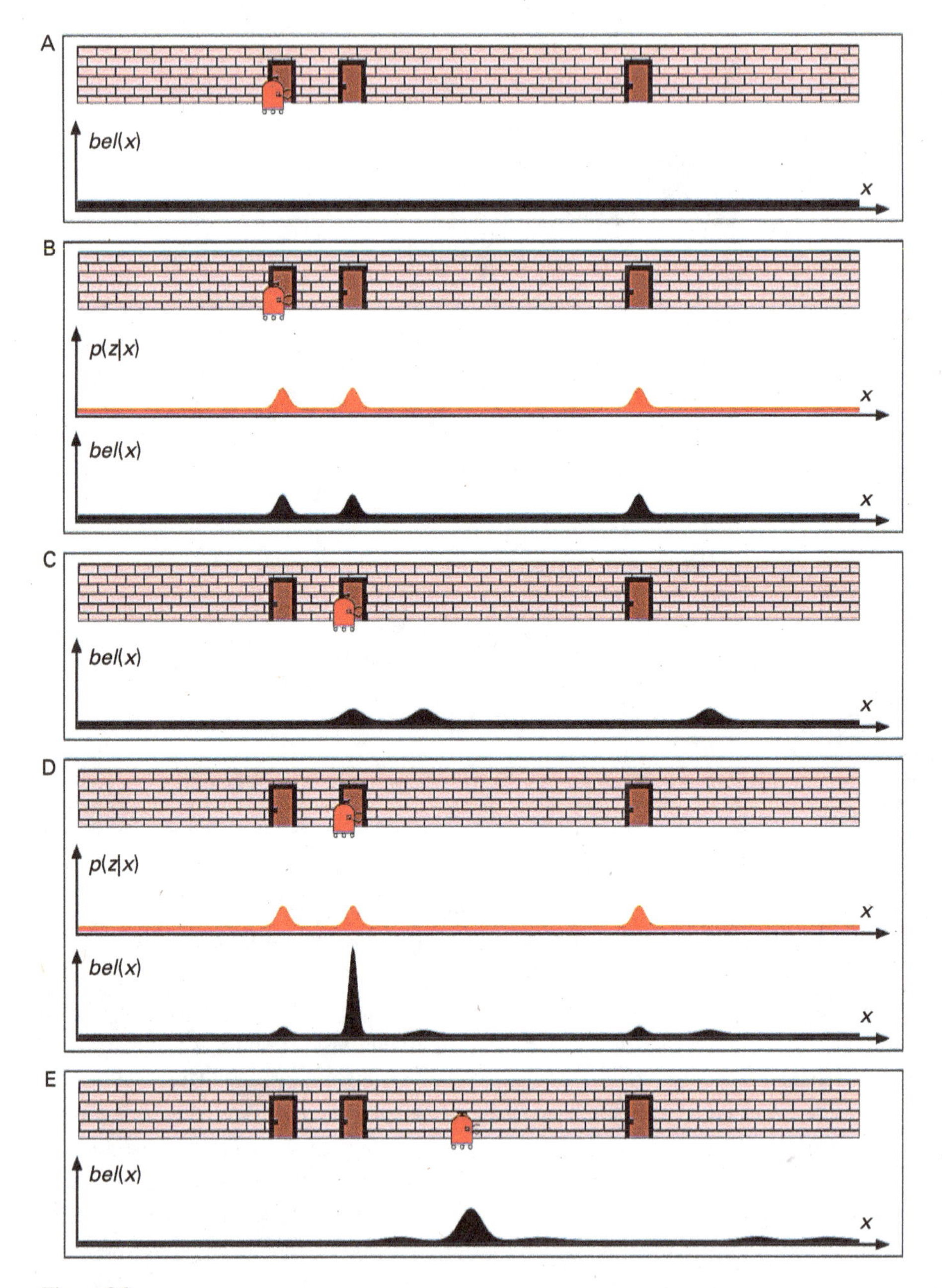

Figure 8.2
Example of Markov localization. The robot has a map of the environment, consisting of a long hallway with multiple doors. *A*, Initially, the robot has a uniform belief over its own position, as it has not sensed or moved. *B*, After sensing the first door, its belief of its position is distributed evenly among all doors in the map. *C*, Moving forward, the belief of the robot's position also moves forward. *D*, Combining the previous belief with sensing the next door, the belief of the position is high around the second. *E*, The belief shifts forward as the robot moves forward. From Thrun et al. (2005).

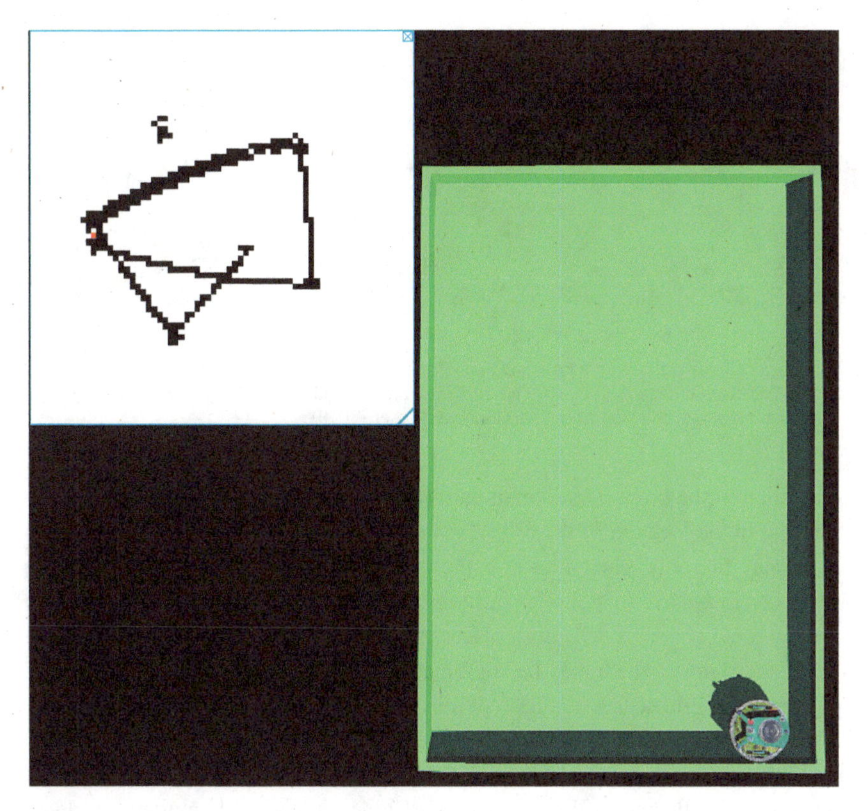

Figure 8.3
Example of error accumulation in robot sensing. Black marks represent the sensed boundaries of the rectangular arena based on the robot's odometry. Image created using the advanced_slam Webots simulation.

to perform dead reckoning for short ranges of space, errors accumulate over time due to sensor noise and slippage. Figure 8.3 shows the occupancy grid of a robot exploring a rectangular arena. The black marks represent the locations where the robot believes there are walls as calculated by its wheel positions (i.e., odometry). Initially, the black marks align well with the arena, but errors accumulate over time due to noisy sensors and wheel slippage. To account for uncertainty in the sensors and controls, probabilistic techniques such as particle filters and extended Kalman filters are often applied in SLAM (Choset et al., 2005).

8.2.2 Neurobiological Representations for Navigation

For every challenge in a mobile robot a biological navigating organism faces a similar challenge. However, biological sensing in a naturalistic environment has many differences from the structured environments of mechanical robots and therefore requires a different set of solutions. One large difference is that the mapping and localization must be achieved using neural circuitry. Experiments using neural recordings of rodents have revealed insights on

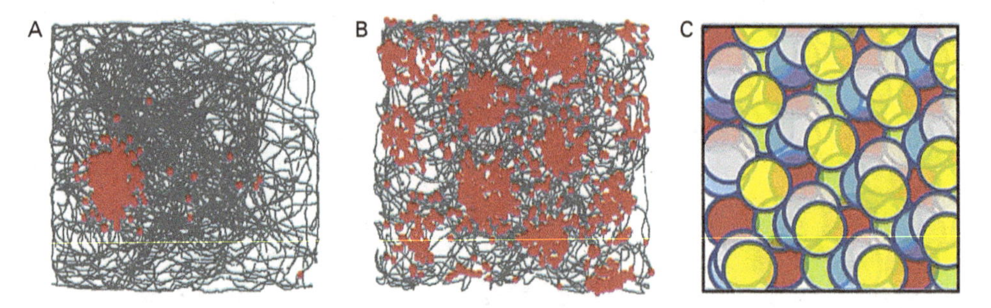

Figure 8.4
A, The firing of a place cell correlates with a rat's position within a room. *B*, The firing of a grid cell correlates with a grid pattern of locations in the room. *C*, The combination of many grid cells and place cells may represent a map of the whole environment. Adapted from Derdikman & Moser (2010).

spatial representation in the brain (Derdikman and Moser, 2010). In the hippocampus, **place cells** fire according to the location of the rodent within a room (figure 8.4*A*). For instance, a single place cell may fire only when a rat is in the bottom-left corner of a room. **Grid cells** in the entorhinal cortex behave in a similar fashion but fire with a repeating gridlike pattern of locations in the room (figure 8.4*B*). Both place cells and grid cells cover a wide range of spatial area sizes and shapes, which may be combined to form a map of an environment. For instance, a population of grid cells with different offsets may combine to cover an entire area. Place cells and grid cells of different sizes represent space at differing resolutions, which are likely recruited differently according to the goals and needs of the rat (see figure 8.4*C*).

In addition to place cells and grid cells, a slew of other spatial and navigation-related cells have been discovered, such as head direction cells (Cullen & Taube, 2017), speed cells (Kropff et al., 2015), goal cells (Hok et al., 2005), border cells (Solstad et al., 2008), and axis-tuned cells (Olson et al., 2017). These neural correlates of space are likely part of multiple neuronal circuits used for navigation. One example of a task involving these cells is **path integration**, which is similar to dead reckoning. Path integration involves tracking past movements to update the representation of one's own location and calculate new paths. Figure 8.5 shows an example of path integration, in which the directions and distances of multiple legs of a journey are integrated to form the calculation of a path back to the starting location. The homing behavior of rats demonstrated path integration, as they are able to remember their past movements and navigate back to a home position, even with the absence of visual input in a dark room (Alyan & McNaughton, 1999). A neural circuit for path integration may include a means of updating one's representation of place using representations of movement and control, such as changes in head direction and velocity.

8.2.3 Cognitive and Semantic Maps

The concept of a **cognitive map** was first introduced by Edward Tolman in the early twentieth century (Tolman, 1948). In his experiments, he made several discoveries regarding how

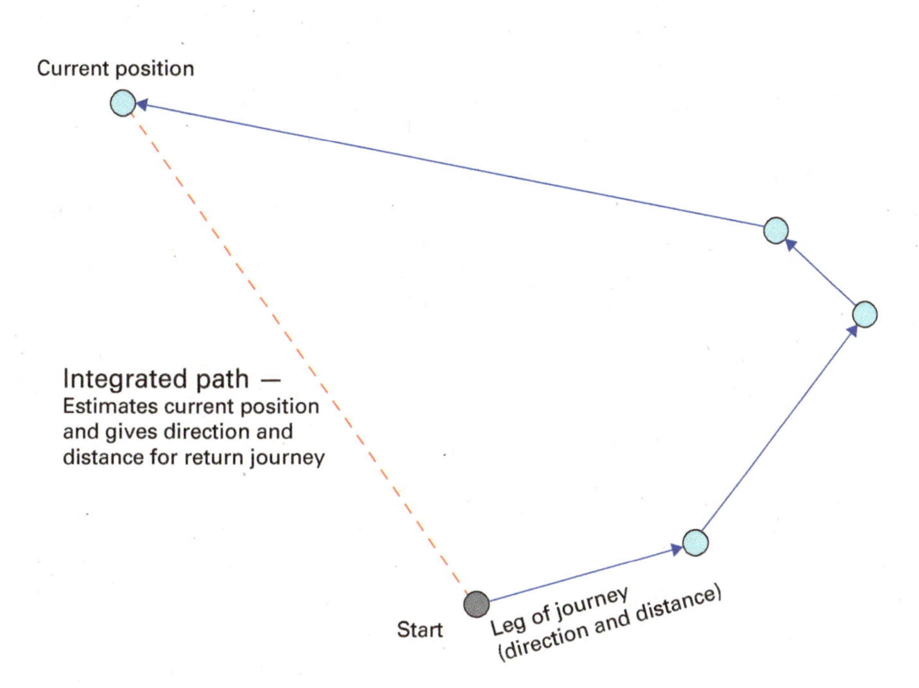

Figure 8.5
Example of path integration. From a starting position, an agent keeps track of past movements as journey legs of specific lengths and distances. This information can be used to formulate a direct homing path back towards a previously visited location. Adapted from https://en.wikipedia.org/wiki/Path_integration.

rats represent space. Cognitive maps express (1) latent learning—acquiring knowledge and leveraging previous learned knowledge for future use even if the payoff may come much later; (2) vicarious trial and error—the ability to weigh one's options before taking decisive action; (3) searching for the stimulus—deciding which cues or signals are relevant for a task; (4) hypotheses—deriving tests to adapt to time-varying contexts and environments; and (5) spatial orientation—the ability to reason through space. In a series of seminal behavior experiments, Tolman and his colleagues showed that rats were capable of all five of these characteristics.

These findings suggested that a cognitive map of space exists in the brain (O'Keefe & Nadel, 1978), which was further supported by the discoveries of place cells and grid cells (Derdikman & Moser, 2010). Since then, the concept of cognitive maps has been used by many to describe semantic as well as spatial features.

Aside from representing space, the hippocampus is known for its involvement in memory, particularly in consolidating short-term memories with long-term memory and semantic knowledge (McClelland et al., 1995). Because the hippocampus covers not only spatial representations but also semantic information, some have found it unnecessary to view spatial knowledge as anything more than a subset of the many stimuli similarly encoded in memory

(Eichenbaum & Cohen, 2014). If so, this may explain why random recall through semantic networks in humans resembles foraging patterns in many animals (Abbott et al., 2015) and why representations of value and spatial position overlap. In this view, place cells may store experiences representing space, and there may be other cells in the hippocampus representing nonspatial information. Mental deliberation can then be thought of as a task parallel to spatial navigation.

Some mapping approaches in robotics combine spatial and semantic information to form semantic cognitive maps, associating locations with related knowledge. Figure 8.6 shows an example of a hierarchical semantic cognitive map for indoor robotics (Galindo et al., 2005). Space is represented as a hierarchical clustering of space, with large areas composed of smaller areas. The spatial clusters are anchored to detected objects, such as furniture in a room. The advantage of such a representation is that the robot can learn to make inferences about different types of areas, which may prove useful in exploring unknown areas. For instance, by learning that a particular spatial cluster contains a stove and refrigerator, the robot learns to associate the two objects with each other, forming the concept or schema of a kitchen. In an unfamiliar space, if the robot sees a stove, it may recall its prior belief that a refrigerator is nearby, aiding in mapping and planning.

8.3 Planning

An agent uses its representation of space to plan paths toward goals. Planning may include short time scales, such as reacting to an immediate obstacle, or longer time scales, such as determining the most efficient way to traverse a large space. For shorter time scales, the relationship between perceptual input and motor output is emphasized because reaction time must be fast. For longer time scales, spatial and semantic maps are emphasized because the agent has the time to reason about concepts and possibilities.

8.3.1 Reactive Planning

Reactive planning occurs along a short time scale and involves quick judgements of motion. In navigation, **obstacle avoidance** is one of the most common reactive planning tasks. For static obstacles, the agent should plan a path around obstacles that minimizes risk of collision while potentially accounting for other factors such as trajectory smoothness. For dynamic obstacles, the agent must additionally predict future positions of the obstacles. Traditional robotics has adopted modeling techniques from physics and mathematics for reactive planning. For instance, the construction of artificial potential fields can be used to plan movements that avoid obstacles and reach a goal position (Raheem & Badr, 2017). Figure 8.7 shows potential fields designed to encode the positions of obstacles and a goal location. Vectors in the potential field point away from the obstacles and toward the goal location, such that the agent can plan a motion path by following the vector directions and magnitudes.

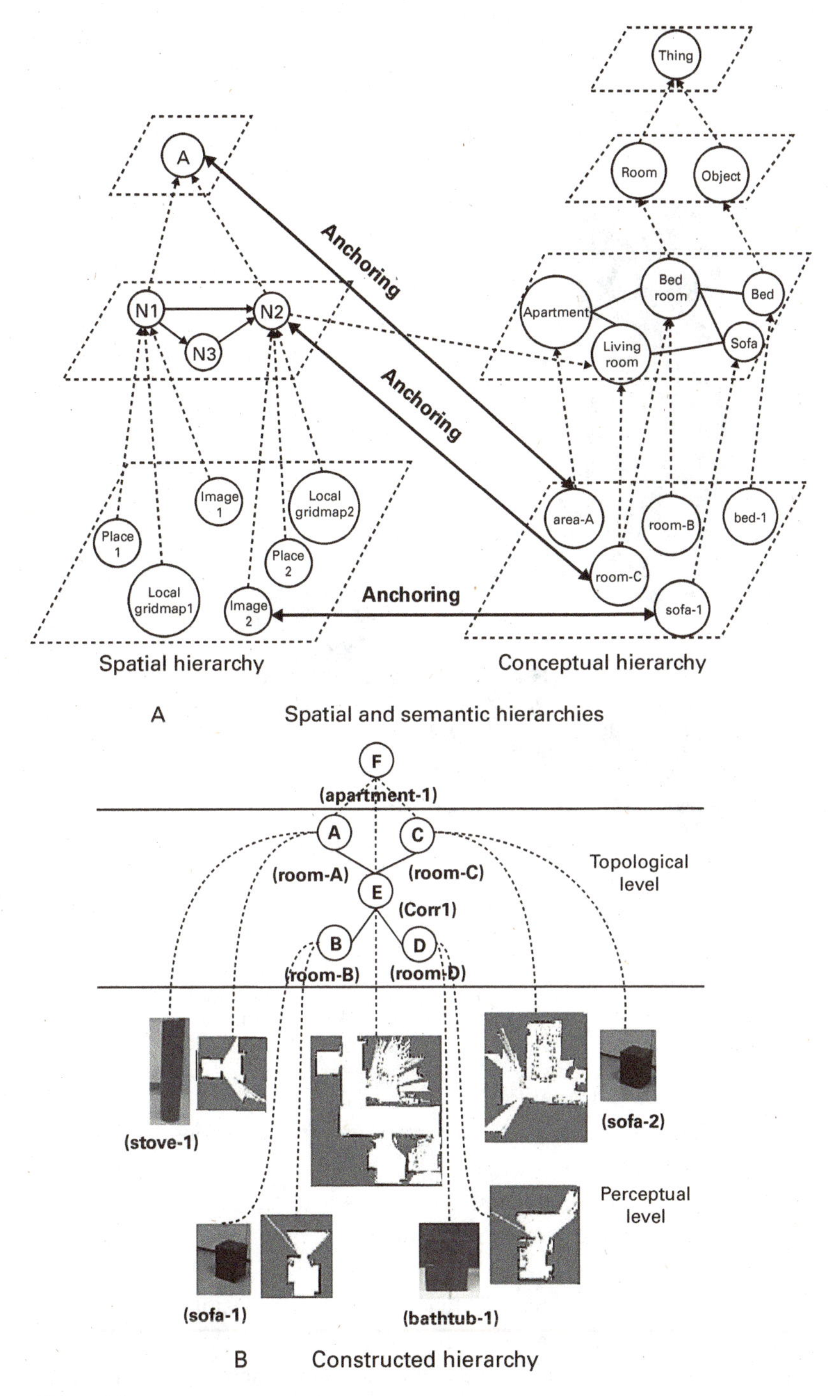

Figure 8.6
A, Anchoring of spatial hierarchy to conceptual hierarchy. The spatial hierarchy consists of perceptual and spatial information, such as images and map portions. The conceptual hierarchy consists of objects, such as furniture on the lower level and rooms on the higher level. Solid lines denote anchoring between concepts in the two hierarchies. Dotted lines refer to inferences, which are symbolic links that can be made. *B*, An example of a hierarchy constructed from robot data. Nodes *A*, *B*, *C*, and *D* associate perceived objects with different parts of an occupancy map. Together, they connect to form the concept of an apartment. Adapted from Galindo et al. (2005).

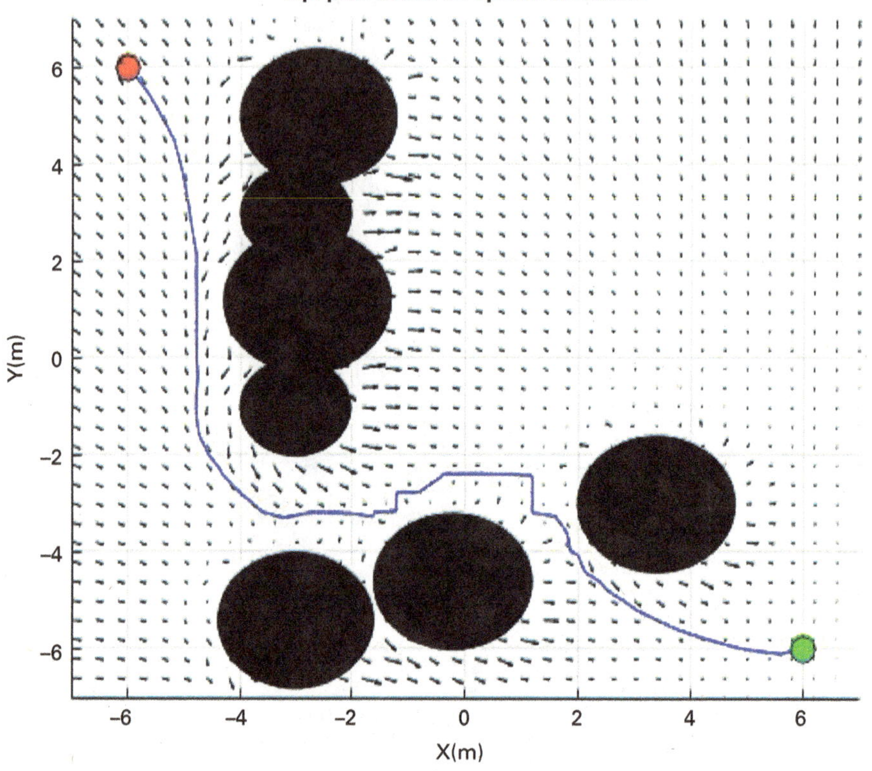

Figure 8.7
Example of using artificial potential fields to plan an optimal path toward a goal destination. The space is represented as a 2D vector field with vectors pointing away from obstacles and toward goal locations. An agent can then use the vector field to create a safe path toward the destination. Adapted from Raheem & Badr (2017).

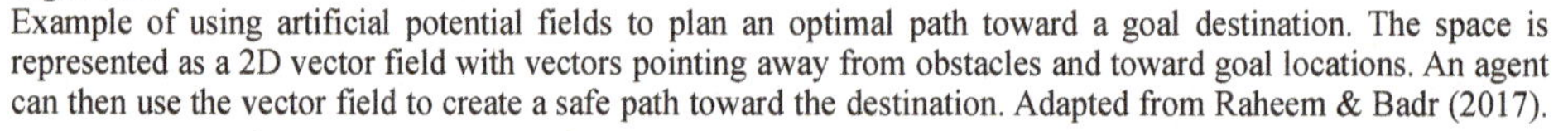

Approaches such as artificial potential fields assume that the locations of obstacles are known. However, in biological systems, the connection between perception and action is intertwined, with each influencing the other simultaneously. In chapter 2, we described a cortical model that closely tied perception and action for reactive control (Beyeler et al., 2015). The model controls an indoor ground robot, taking camera frames from the robot and preprocessing them as input to the neural model. The primary visual cortex extracts optic flow, which projects to cortical area MT that represents the positions and sizes of obstacles using variations in optic flow as the agent moves through space. The robot used this information to steer away from obstacles and navigate towards a target. The trajectories of the robot were very similar to a dynamical systems model of human obstacle avoidance and steering (Fajen and Warren, 2003). By linking first-person perception of motion directly to a steering action, the resulting behavior is likely to differ from the traditional approach of calculating trajectories based on third-person perspectives.

8.3.2 Goal-Driven Path Planning

Goal-driven path planning consists of the selection of waypoints to reach an end goal. When the scale of waypoints is small, the path planning is similar to reactive planning. Large-scale path planning may not account for dynamic obstacles and is more concerned with the overall strategy. Dijkstra's algorithm is a well-known approach for finding the shortest paths. Figure 8.8 depicts the use of the algorithm in finding the shortest path from a start location to a goal location in a grid containing an obstacle. The algorithm is guaranteed to find the shortest path. However, it can be computationally intensive to explore all nodes in the space. The A-star path planning algorithm, also denoted as A*, is a faster variant of Dijkstra's algorithm that selects the next nodes to explore using heuristics, that is, general rules. This leads to finding the shortest path with less computation and is guaranteed to find the shortest path if the heuristic meets a set of requirements.

In the next section we describe a case study of a form of Dijkstra's algorithm implemented with a spiking neural network.

8.4 Case Study: Spiking Wavefront Propagation

In this case study we describe a neuromorphic path planning algorithm, implementing a form of Dijkstra's algorithm using spiking dynamics. As mentioned in previous chapters, neuromorphic hardware is designed to save energy by performing similar computations as neurons in the brain. Unlike conventional computers neuromorphic hardware runs in an asynchronous, event-driven fashion by incorporating small computational units that communicate sparsely through spiking. The result is an energy-efficient form of computation that could potentially perform brainlike functions. In use cases such as robotics the power

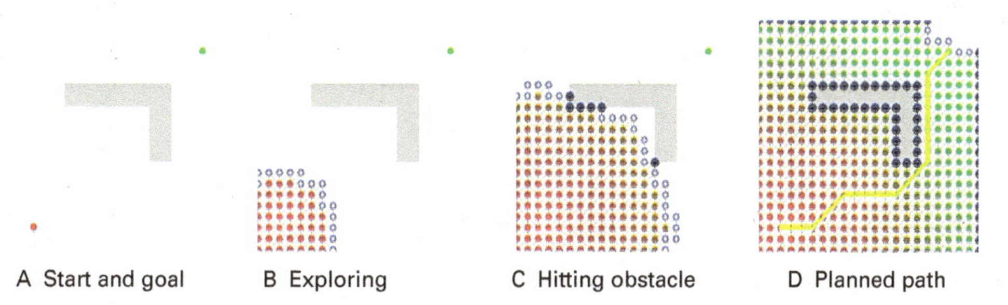

Figure 8.8
Steps of Dijkstra's algorithm. *A*, The red and green dots mark the start and goal locations, respectively. *B*, A set of explored nodes in the space is built up, expanding outward from the start location. The color of the nodes shows the distance of the shortest found path from the start location to that node. Nodes marked by empty circles show the next nodes to be explored. *C*, The explored nodes hit an obstacle. Because those nodes are unable to travel further, they are marked in black. *D*, The final planned path to the destination is extracted by tracing backward from the destination, picking nodes that have the shortest distance to the start. Adapted from https://commons.wikimedia.org/wiki/File:Dijkstras_progress_animation.gif.

savings are particularly beneficial because energy is limited. Accompanying neuromorphic hardware are neuromorphic algorithms, which are inspired by neuroscience. This is where neuromorphic engineering and neurobiology can combine for better understanding of both fields.

Planning routes and paths are behaviors that both animals and robots must perform under environmental constraints such as limited time and energy. A class of path planners that can find optimal paths includes wavefront planners, potential fields, and diffusion algorithms (Barraquand et al., 1991; Soulignac, 2011). Typically the algorithm starts by assigning a small number value to the goal location. In the next step the adjacent vertices (in a topological map) or the adjacent cells (in a grid map) are assigned the goal value plus one. The wave propagates by incrementing the values of subsequent adjacent map locations until the starting point is reached. The wave cannot propagate through obstacles. A nearly optimal path, in terms of distance and cost of traversal, can be read out by following the lowest values from starting location to the goal location.

Spiking wavefront propagation is a neuromorphic navigation algorithm inspired by neuronal dynamics and connectivity of neurons in the brain (Hwu et al., 2018; Krichmar, 2016). The algorithm is loosely based on the dynamic responses of place cells in the hippocampus. In hippocampal preplay, place cells activate in sequence according to future trajectories that may be taken by the animal (Dragoi & Tonegawa, 2011; Pfeiffer & Foster, 2013). Spiking wavefront propagation is also supported by the observation that spreading activation of brain activity is seen in several areas of the brain including the hippocampus and that this is involved in memory systems (Zhang and Jacobs, 2015).

With these neurobiological phenomena in mind, a spike-based model of wavefront planning can be constructed (Hwu et al., 2018; Krichmar, 2016). The spiking wavefront propagation algorithm assumed a grid representation of space. Each grid unit corresponded to a discretized area of physical space, and connections between units represented the ability to travel from one grid location to a neighboring location. Each unit in the grid represents a single neuron with spiking dynamics, which are captured with a model described by the following equations.

The membrane potential of neuron i at time $t+1$ is

$$v_i(t+1) = u_i(t) + I_i(t), \tag{8.1}$$

in which $u_i(t)$ is the recovery variable, and $I_i(t)$ is the input current at time t. The recovery variable $u_i(t+1)$ is described in the following equation:

$$u_i(t+1) = \begin{cases} -5 \textit{ if } v_i(t) = 1; \\ \min(u_i(t)+1, 0) \text{ otherwise} \end{cases} \tag{8.2}$$

such that immediately after a membrane potential spike, the recovery rate starts as a negative value and linearly increases toward a baseline value of 0.

The input current I at time $t+1$ is

$$I_i(t+1) = \sum_j \begin{cases} 1 \; if \; d_{ij} = 1 \\ 0 \text{ otherwise} \end{cases} \tag{8.3}$$

such that $d_{ij}(t)$ is the delay counter of the signal from neighboring neuron j to neuron i. This delay is

$$d_{ij}(t+1) = \begin{cases} D_{ij}(t) \; if \; v_j = 1 \\ \max(d_{ij}(t) - 1, 0) \text{ otherwise,} \end{cases} \tag{8.4}$$

which behaves as a timer corresponding to an axonal delay with a starting value of $D_{ij}(t)$, counting down until it reaches 0. The value of $D_{ij}(t)$ depends on the environmental cost associated with the spatial area covered by the neuron. Input currents come from neighboring connected neurons. When a neighboring neuron spikes, the input current I_i is set to 1. This triggers the neuron to spike because its voltage reaches 1. After the spike the recovery variable u_i is set to −5 and then gradually returns to 0, modeling the refractory period. Further, immediately after spiking, all delay counters d_{ij} for all neighbor neurons j are set to their assigned starting values of D_{ij}. A diagram of the voltage of a spiking neuron is shown in figure 8.9.

The D_{ij} values represent the environmental cost of traveling from neuron j to neuron i and can be defined using a number of strategies. The cost can include factors such as the energy required for traversal, the roughness of the terrain, and potential risks. If these factors are known in advance, a cost map of the environment can be formed as a grid with units

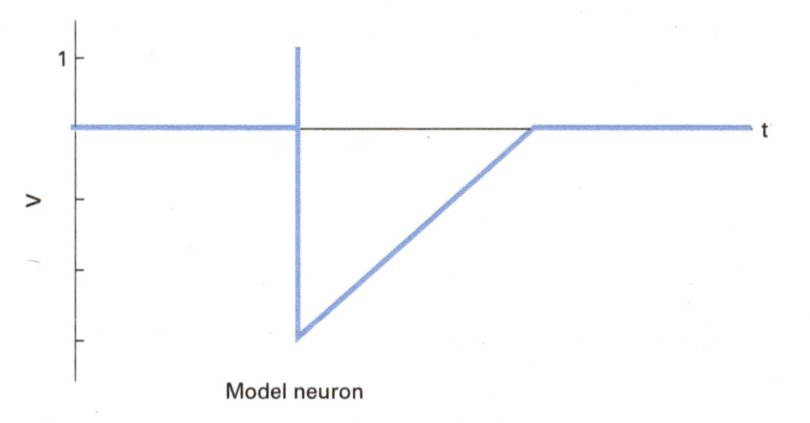

Figure 8.9
The voltage of a neuron starts at 0. When incoming current from a neighboring neuron arrives after the set delay, the voltage is set to 1, indicating a spike. Immediately after the spike the recovery variable induces the voltage to be a negative state and then gradually returns to 0.

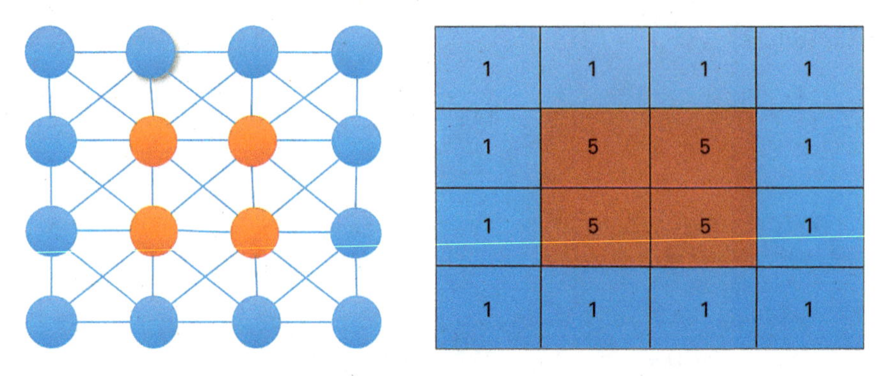

Figure 8.10
Axonal delays corresponding to a cost map of the environment. Red-colored neurons have high delay values, such that spikes from neighboring neurons take a long time to arrive, whereas the blue-colored neurons have small delay. The values of axonal delays can be initialized using a cost map, as shown on the right.

matching the neuron arrangement. Figure 8.10 shows an example of a cost map with a high cost in the center and low cost on the edges.

If a cost map of the environment is available, the cost related to each spatial unit can be directly translated into delay values. There is also the option of learning costs through experience as the agent explores the area. This can be done by starting with a uniform cost or delay value across the entire area and then updating it during exploration. For areas of high cost the delay is increased, and vice versa for areas of low cost.

Learning in the spike wavefront algorithm was inspired by evidence suggesting that the myelin sheath, which wraps around and insulates axons, may undergo a form of activity-dependent plasticity (Fields, 2015). These studies have shown that the myelin sheath becomes thicker with learning motor skills and cognitive tasks. A thicker myelin sheath implies faster conduction velocities and improved synchrony between neurons.

These values of D_{ij} are updated each time the agent visits a location in the grid using the following equation:

$$D_{ij}(t+1) = D_{ij}(t) + \delta(\text{map}_{xy} - D_{ij}(t)), \tag{8.5}$$

where δ represents the learning rate and map_{xy} represents the observed cost for traversing the location (x, y), which corresponds to neuron i. This rule is applied for each of the neighboring neurons, j, of neuron i. By using this method of axonal plasticity the agent can simultaneously explore and learn, adapting to changes in the environment. Learning the costs of the environment in this manner may take many trials of exploration. Therefore, it may be easier to start with a prior cost map of the environment and update with new values as they are observed. When the learning rate is sufficiently small, the updating rule can account for small sampling errors due to sensors or random environmental fluctuations as the updating is averaged over multiple trials.

To use the path-planning algorithm with delay-encoded costs, the neuron corresponding to the current location is induced to spike. This causes an input current to be sent to neighboring neurons, starting a traveling wave of activity across the area covered by the grid. As each neuron spikes, the spike index and the time of spike are logged in a table. This is known as address event representation (AER) in neuromorphic computing. Figure 8.11 illustrates how AER information is used to trace a path from start to goal.

To extract a path from the AER, a list of neuron IDs is maintained, starting with the goal neuron. The first spike time of the goal neuron is found. Then the timestamps are decremented until the spiking of a neuron neighboring the goal neuron is found. That neuron is then added to the list. The process then continues by finding spikes of neurons neighboring the most recent neuron added to the list until the start neuron is added. The result is the optimal path between the start and goal in reversed order.

The spiking wavefront algorithm was tested successfully on an outdoor ground robot across a large park with terrains including grass, asphalt, and dirt (Hwu et al., 2018). The ground robot was an Android-based robotics platform (Oros & Krichmar, 2013) consisting of off-the-shelf commercial parts and an Android phone running the spiking wavefront planning algorithm. The actuators and sensors of the robot were powered by a nickel metal hydride (NiMH) battery. Two areas of the park were used for testing. One of the areas consisted of a large grassy area surrounded by an asphalt trail. From this area, three cost maps were generated (see figure 8.12): (1) a uniformly low cost throughout; (2) a low cost for the asphalt road and medium cost for the grass; and (3) a low cost for the road, medium cost for the grass, and high cost for park benches and other obstacles. The second area had a dirt trail

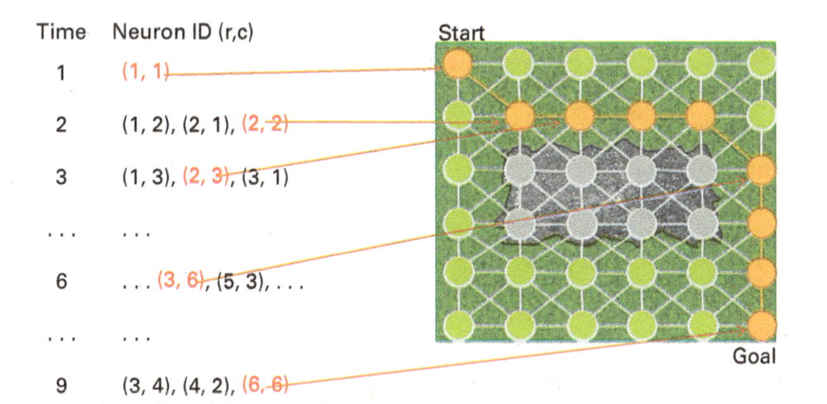

Figure 8.11
Example of using address event representation (AER) for extracting a path. Information includes time stamps with IDs of neurons that spiked on that time stamp. After a spike wave has propagated from start to goal a path can be traced by maintaining a list of neuron IDs, starting with the goal neuron. The earliest spike time of the goal neuron is found in the address event representation, and then the earliest spike time of neurons neighboring the goal neuron is found and that neuron's ID is added to the list. This process continues until the start neuron is reached, at which point the path can be obtained by reading out the list of neuron IDs.

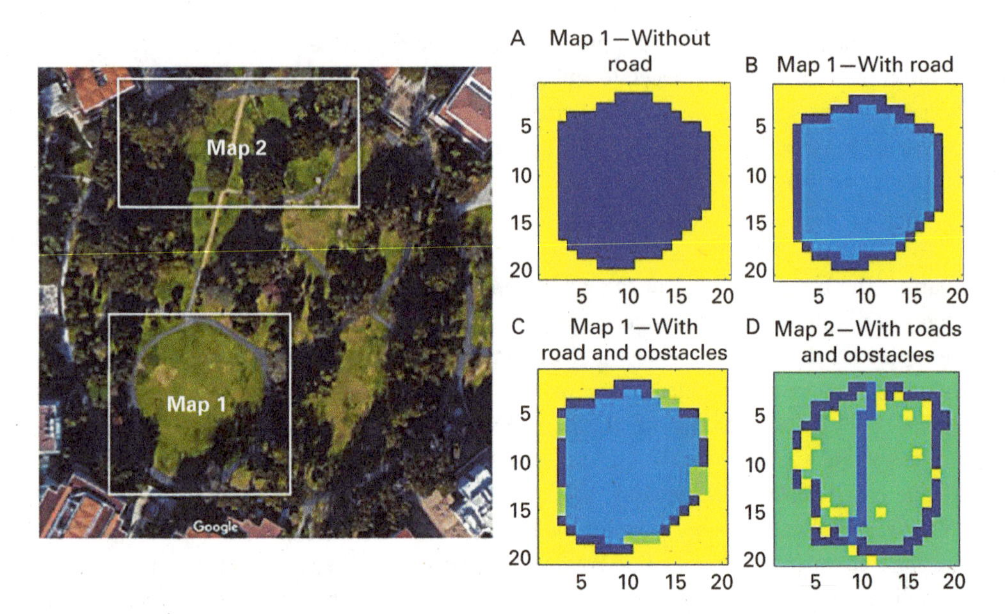

Figure 8.12
Cost maps generated from a real outdoor environment for testing spiking wavefront propagation. *Left*: An overhead satellite view of a grassy environment from which cost maps were generated. *Right*: Different cost maps. *A–C*, Map 1 cost map. *A*, The entire area was assigned a low cost. *B*, The road received a low cost, and the grass received a medium cost. *C*, The road had a low cost, and the park benches had a medium cost. *D*, Map 2 cost map. The road had a low cost, the trees had a high cost, the dirt path had a medium cost, and the grass had a medium-high cost.

cutting across the grassy field and scattered trees. A cost map was created for this area, with a low cost for the asphalt, a medium cost for the dirt road, a medium-high cost for the grass, and a high cost for trees. For each of these cost maps grids of neurons were created, with delays according to the costs. A set of start and end points were selected and run on each map using spiking wavefront propagation.

On the basis of the cost map different routes were planned with the same sets of start and goal positions (see figure 8.13). This demonstrated a trade-off between taking the shortest route over rough terrain and taking a longer path by avoiding high-cost terrains. For the cost map with uniform cost, the shortest route was chosen every time. For the cost map with low cost for the asphalt road, the paths tended to stay on the road. For the cost map with the dirt path, the robot would occasionally choose to travel on the dirt path. The experiments showed that spiking wavefront propagation could be performed on an energy-restricted robot, outputting traversable paths that handled energy trade-offs between taking the shortest path and choosing smoother terrains.

The spiking wavefront path planning algorithm is compatible with neuromorphic hardware, as shown by implementing it on the IBM TrueNorth NS1e (Fischl et al., 2017). The

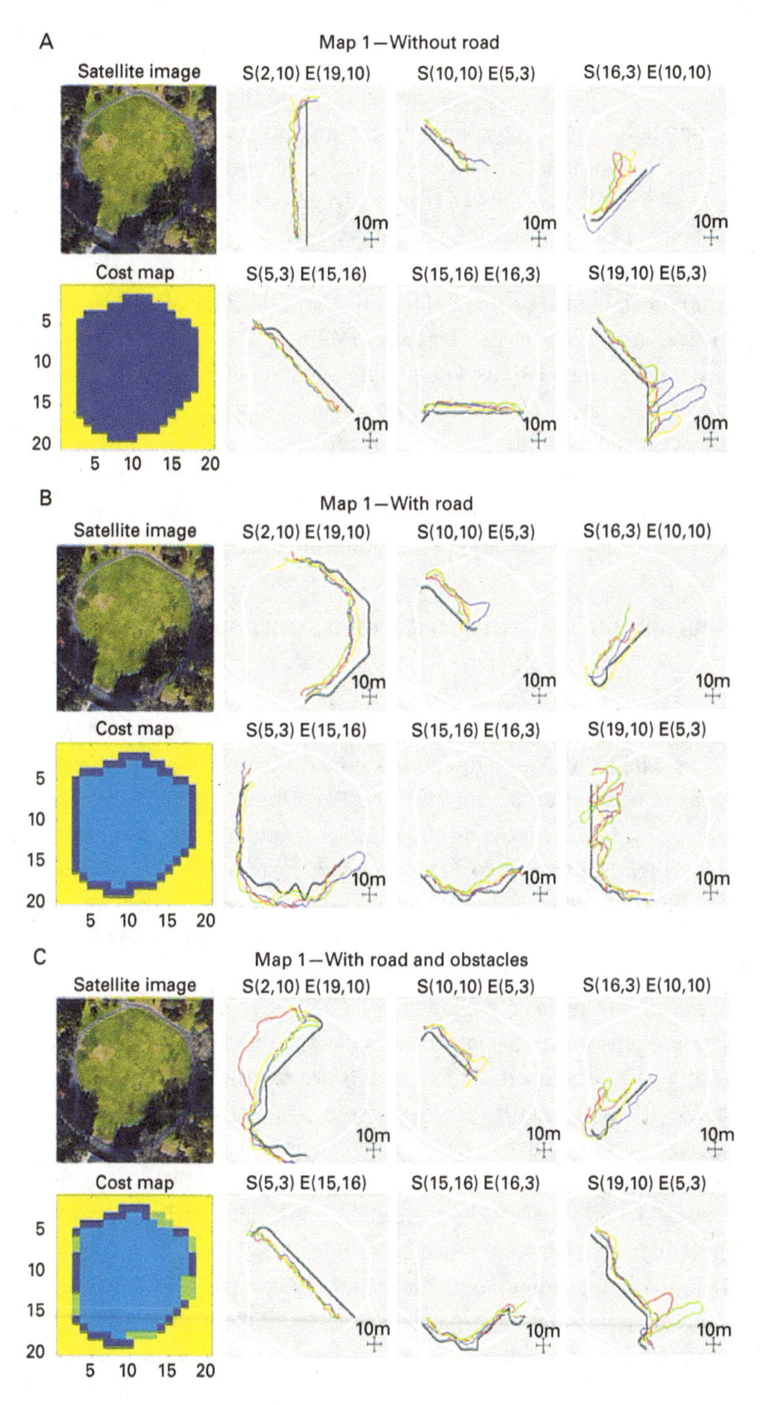

Figure 8.13
Experimental results for the spiking wavefront path planning algorithm on map 1. *Left*: Columns show three different cost maps. *A*, High cost for boundary and uniform cost for the rest of the area. *B*, Same as *A* except lower cost for the paved path than for the grassy area in the middle. *C*, Same as *B* except higher costs for benches and plants near the paved path. Black lines indicate the planned route, and colored lines indicate the actual route taken by the robot. *Right*: Charts in *A–C*. Different trajectories taken by the robot are shown by colored lines, and the desired trajectory from the spiking wavefront algorithm is shown in black.

TrueNorth's extremely parallel computing coupled with low-power neuronal elements demonstrated large efficiency in computing speed and energy consumption.

A separate neurorobot experiment also showed that the TrueNorth NS1e was able to run on the Android-based robotics platform, sharing the battery used to power the robot. The TrueNorth chip took camera images from the robot and trained a convolutional neural network (CNN) to steer left or right along a mountain path (Hwu, Isbell, et al., 2017). After the CNN was trained it was deployed on the TrueNorth chip that was mounted on the robot. The robot traversed these mountain trails with minimal intervention for nearly an hour on the robot's NIMH battery.

Taken together, these experiments show that outdoor navigation can be implemented entirely on neuromorphic hardware for energy savings (Hwu, Krichmar, and Zou, 2017). The energy efficiency allows more computation to be performed on the same robots and tests neuroscience principles of how the brain operates on limited resources.

8.5 Case Study: Neurobiologically Inspired Robot Navigation and Planning

We have seen that navigation consists of many functions working together, from perception to planning and action. The next case study models how different brain areas interact to perform these functions (Cuperlier et al., 2007). Cuperlier, Quoy, and Gaussier developed a model of navigation based on the learning and representation of transitions between places. The challenge for their study was to (1) develop a system that autonomously decides the appropriate behavior to achieve a goal; (2) use biologically grounded methods; (3) use sensing found in rodents (e.g., visual information but not GPS or lidar); and (4) avoid the tendency of models to require an external algorithm to perform actions.

To meet these challenges and constraints, they closely examined the role of the hippocampus in navigation. Hippocampal lesions impair navigation activities, and hippocampal place cells fire when the animal is at a particular location. Place cells are rapidly recruited in a new environment, and the same cell can fire in different locations in different environments. Place cells are stable over time and multimodal; that is, they respond reliably in the dark and therefore do not rely only on visual input. Place cells are dependent on proximal and distal landmarks.

Capturing these biological details, their model merges the appearance of landmarks (the *what*) and their azimuths (the *where*) to form place cells. Place cell activities are input to the dentate gyrus (DG) in the hippocampus. The model introduced the idea of transition cells in the CA area of the hippocampus. These cells learn transitions between places as the agent travels. The connections between transition cells are then learned and stored in a cognitive map, which can be used for path planning. Figure 8.14 shows the neural network architecture for their model.

The model was tested in simulations and on a physical robot. Figure 8.15 shows how landmarks are extracted from the visual input of a robot. A difference of Gaussians (DoG)

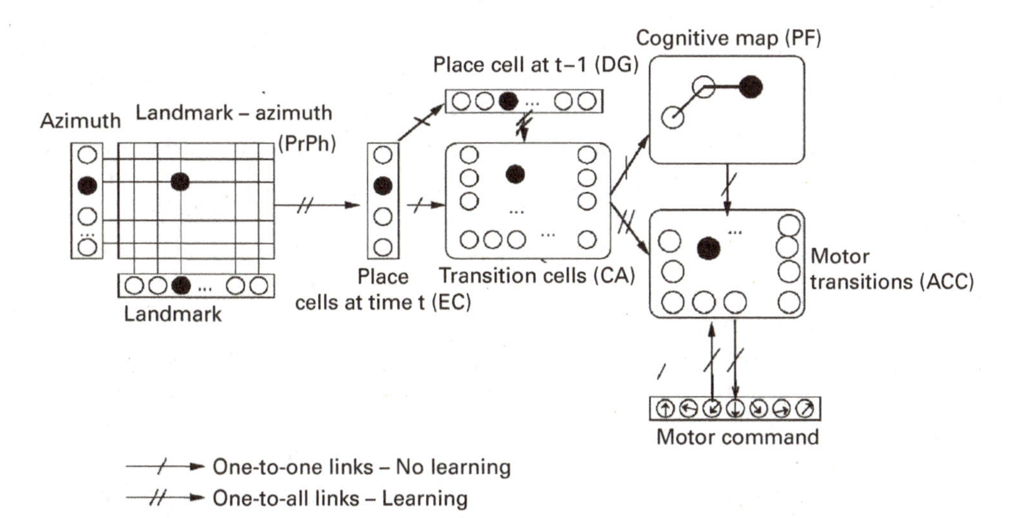

Figure 8.14
Model overview. Conjunctions between landmarks and azimuths are stored in the perirhinal cortex and parahippocampal cortex (PrPh) and then represented as place cells in the entorhinal cortex (EC). Transition cells (CA) then learn conjunctions of place cell activity from the current time step (EC) and previous time step (DG). The cognitive map (PF) then links transition cells. The motor transitions (ACC) then interrogate the transitions and cognitive map with motor commands. Adapted from Cuperlier et al. (2007).

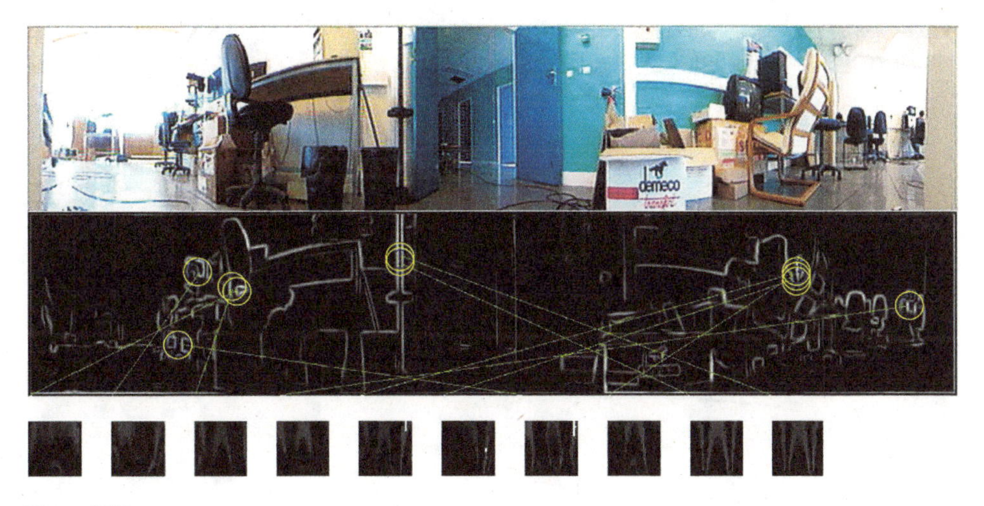

Figure 8.15
Extraction of landmark features. *Top*: Camera input of panoramic view. *Center*: Difference of Gaussians (DoG) filter applied to camera input. Circles show distinct visual landmark features extracted from the filtered image. *Bottom*: Examples of features after log-polar transform, which are then used in the network model. Adapted from Cuperlier et al. (2007).

filter was applied to an image from the panoramic view, extracting small subimages that serve as visual landmarks. The landmarks and their azimuths (angular direction with respect to the agent's heading) were sufficient for place cell recognition.

Similar to the wave propagation algorithms described previously, the model was capable of planning paths by activating a goal neuron and then diffusing activity to the robot's location. Figure 8.16 shows how the robot plans a path from location B to location D. A motivation signal activates all transition cells that transition to location D. The activity is diffused across the transition cells until it reaches a transition cell that is transitioning from B. The highest transition cell activity, in this case BC, is selected, and the robot will then move to location C. This process is continued until the robot reaches its destination. Figure 8.17 shows the place cells and cognitive map learned by a robot navigating an office space.

Their model was able to learn a cognitive map and successfully navigated both indoors and outdoors on different robots. Compared with other biologically plausible models which rely primarily on place cell representations, this model relied on a combination of place cells and transitions between place cells. One advantage of this approach is that it inherently handles environments in which multiple routes use the same place. The model provides a unified vision of the spatial and temporal functions of the hippocampus, combining navigation and memory into a single model. It also overcomes shortcomings of other place cell models by introducing transition cells, with the choice of movement triggered by the activations of transitions rather than places. Recent evidence supports the existence of transition cells, a testament to the predictive power and usefulness of testing models of neurobiology under realistic constraints.

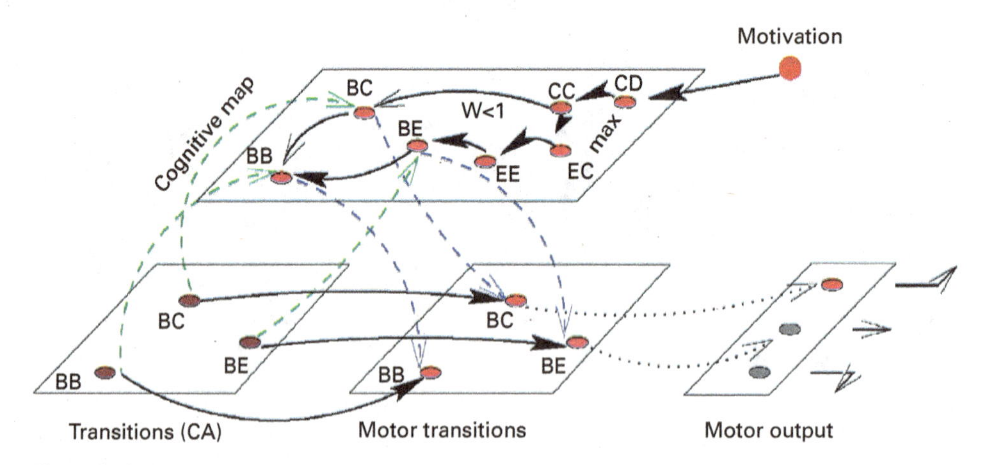

Figure 8.16
Linkages between transitions, the cognitive map, and motor transitions, and their use in path planning. An external motivation value activates a destination in the cognitive map, which is CD in this example. Transitions ending in D are then activated and then continue to diffuse throughout the entire map. The cognitive map biases the activations of motor transitions, which in turn influence motor output. Adapted from Cuperlier et al. (2007).

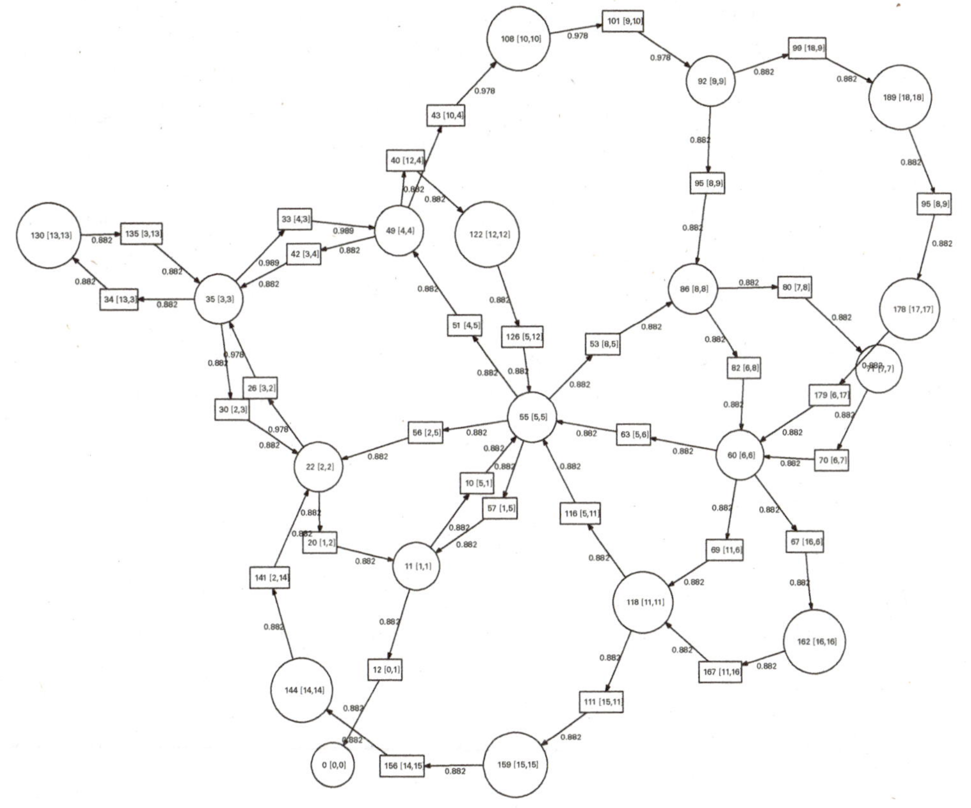

Figure 8.17

Top: Place fields formed during partial exploration of a space. Each place field is marked by a separate color. *Bottom*: Cognitive map after exploration, showing transition cell locations. Large circles represent self-referential transitions between one place cell and itself, and rectangles represent transitions between two different place cells. Adapted from Cuperlier et al. (2007).

8.6 Case Study: RatSLAM—an Application Oriented Model of Rodent Navigation

RatSLAM is a brain-inspired model of navigation that takes a more practical approach than the previous case study (Ball et al., 2013; Milford et al., 2016). RatSLAM focuses on the application of the model to improve upon state-of-the-art SLAM methods used in engineering. Conventional methods were used whenever they provided an advantage. Therefore, they included solutions whenever the underlying biology did not meet their robot's needs.

Figure 8.18 provides an overview of RatSLAM, which is based on computational models of the navigational processes in the hippocampus. The system consists of three major modules: pose cells, local view cells, and an experience map. Pose cells are a set of continuous attractor network (CAN) units. In an attractor network, the activity of the network settles into a basin of attraction, which is represented by a population of active neurons. CAN units are connected by excitatory and inhibitory connections such that stimuli are encoded as stable clusters of activated units, referred to as activity packets or energy packets. Attractor networks, a popular modeling technique for neuron populations, show similar characteristics to grid cells for navigation. In RatSLAM, the centroid of the activity packet encodes an agent's best internal estimate of the current pose. The packet moves on the basis of an estimate of the robot's movement via odometry.

The inputs of the pose cell attractor network come from the activity of local view cells. Local view cells are an expandable array of units, each of which represents a distinct visual scene in the environment. Figure 8.19 shows an example of how local view cells encode information from raw visual input. When a novel visual scene is perceived, a new local view cell is created and associated with the raw pixel data in that scene. When that view is seen again, the local view cell is activated, which then injects activity into the corresponding pose cells.

By combining the view cell and pose cell activities, an experience map is created. An experience map is a graph that uses actions performed by the agent to create nodes or links between poses, which are represented as nodes and edges in a graph. Each experience has an associated position and orientation, and creating a new node also creates a link to the previously active node. When nodes are revisited, there is sometimes a mismatch between perceived positions of the second time versus the first time, due to errors in odometry. The graph is then "relaxed" to fix the error.

Figure 8.20 shows an example of the rat-sized iRat robot navigating *Australia land*. The figure shows the view from the robot's camera and how the experience map was formed. Note that the experience of revisiting places caused the experience map to become more accurate over time (figure 8.20, center) and view cells to be reused (compare experience ID to the view cell template ID shown in figure 8.20).

The RatSLAM method was tested in two realistic applications, each showing the model's effectiveness over state-of-the-art approaches at that time. In one test, a Pioneer P3-DX robot

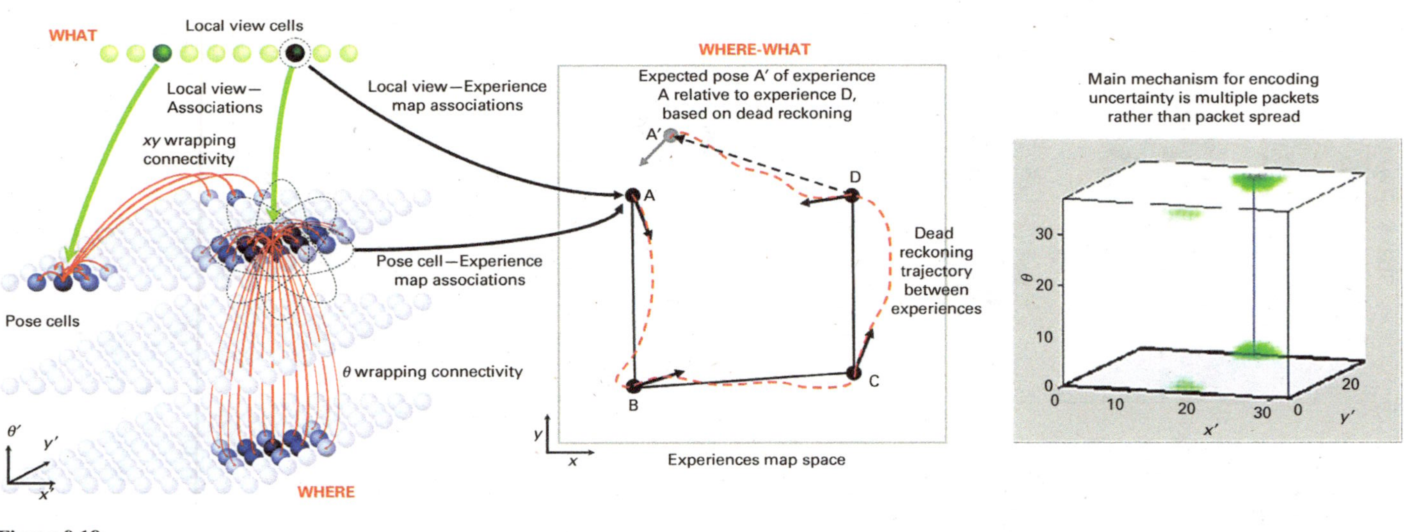

Figure 8.18
Overview of RatSLAM model. *Left*: Local view cells feed into a population of pose cells, which code the x, y, and head direction of the agent as an activity packet. *Center*: Linked-together pose cells form an experience map of the environment. If a place is revisited, the graph is adjusted to fix errors in odometry. *Right*: To represent uncertainty, multiple packets encode potential candidates of the agent's location. Adapted from Milford et al. (2016).

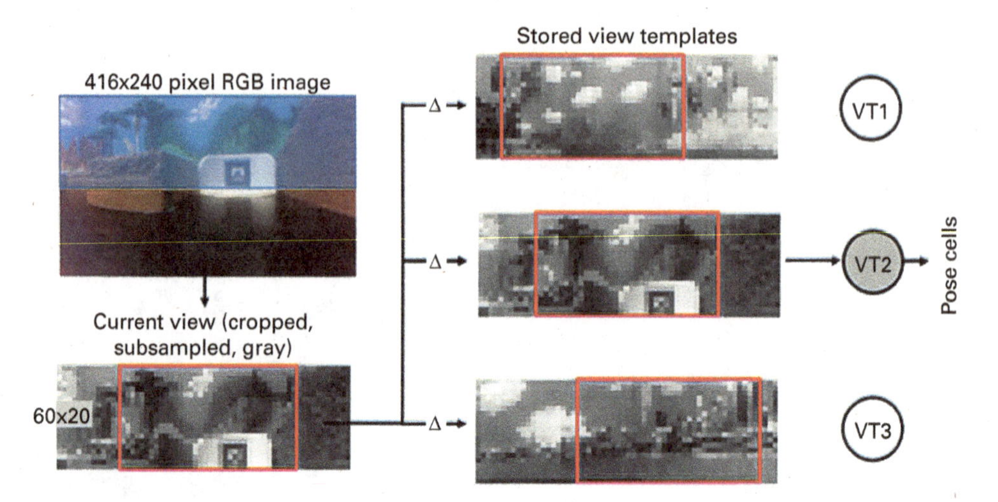

Figure 8.19
Processing of a camera image to a local view cell. A raw RGB image is cropped, subsampled, and converted to grayscale. Subsections of these processed images are encoded as local view cells, which then feed into pose cells. Adapted from Ball et al. (2013).

explored an indoor office environment. The robot started in an unknown office and laboratory complex, autonomously exploring the complex for two hours. Delivery locations were assigned, and 1,000 deliveries were made over a two-week period. The robot maintained its batteries by locating and docking with its charger. Mapping and navigation were continual processes for the entirety of the experiment. The robot was able to form an accurate map of the complex environment. Moreover, it performed well at the "kidnapping" problem, which involves picking up the robot and moving it to an unknown area. The robot was able to learn the unknown area and was still able to recognize previously learned areas when kidnapped back to previously visited places. The same set of cells was used to map both buildings, with little interference in the representations.

Another application of RatSLAM was the mapping of a suburb of Brisbane, Australia (see figure 8.21). A laptop with a webcam was mounted on top of a car. The webcam provided input for both the local view and for visual odometry. An area covering 66 km was mapped over two hours. Forward speed of the vehicle was estimated from the change in scanline intensity profile and rotated previous profile of the camera images. The resulting RatSLAM map is shown in figure 8.21. Note how closely the experience map matches the road map. The experiment showed that RatSLAM could successfully map a large naturalistic environment. Such demonstrations showed that the model was robust to variations in lighting, dynamic environments, and other sensor noise.

RatSLAM showed that a biologically inspired mapping system could compete with or surpass the performance of conventional probabilistic robot mapping systems. It resulted in vision-based navigation that could be achieved at any time of day or night, during any

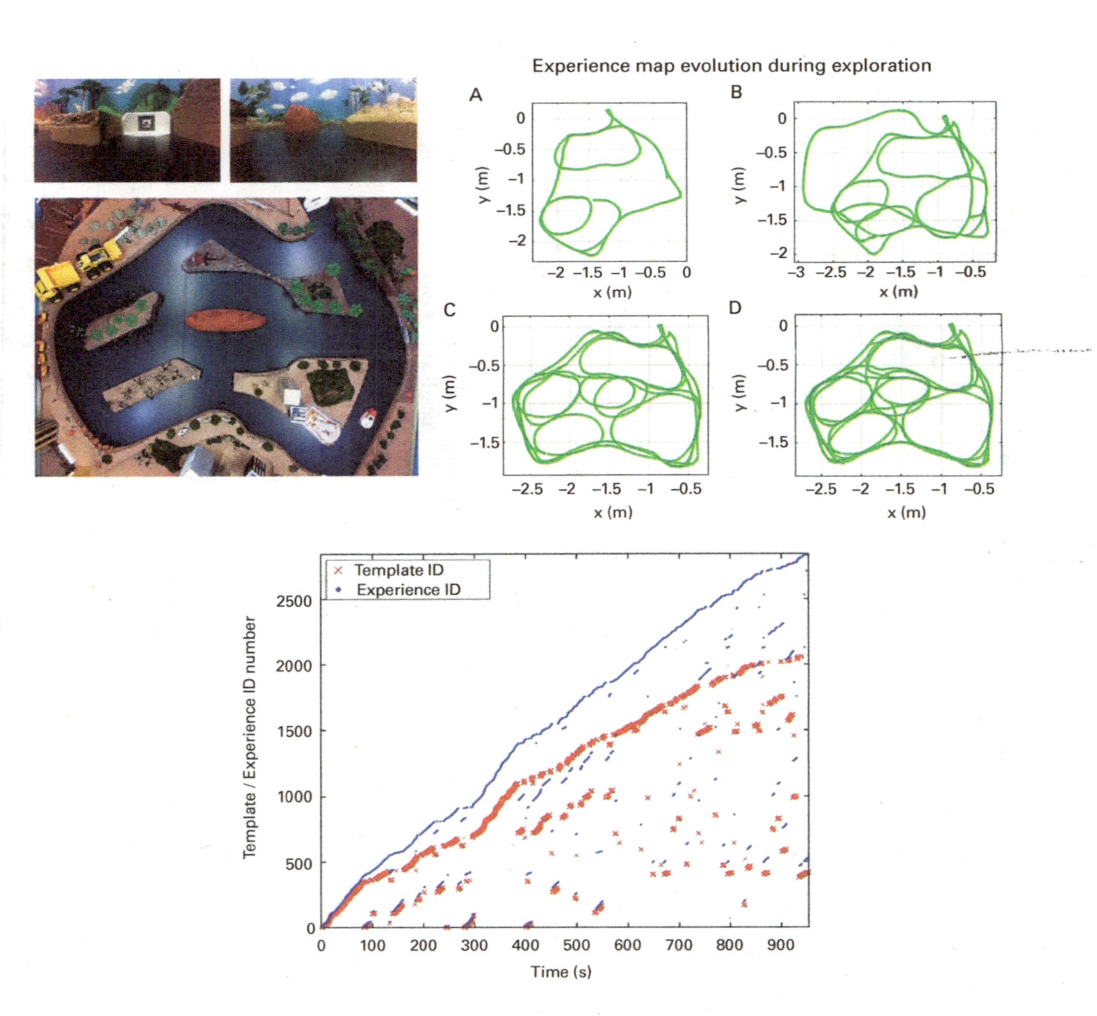

Figure 8.20
Robotic experiment of RatSLAM. *Top left*: Small roadway environment in the shape of Australia. *Top right*: Emergent experience maps. Odometry errors are fixed over time, resulting in an accurate map. *Bottom*: Accrual of new experiences increases over time. The number of templates of unique experiences also increases over time but at a slower rate as places are revisited. Adapted from Ball et al. (2013).

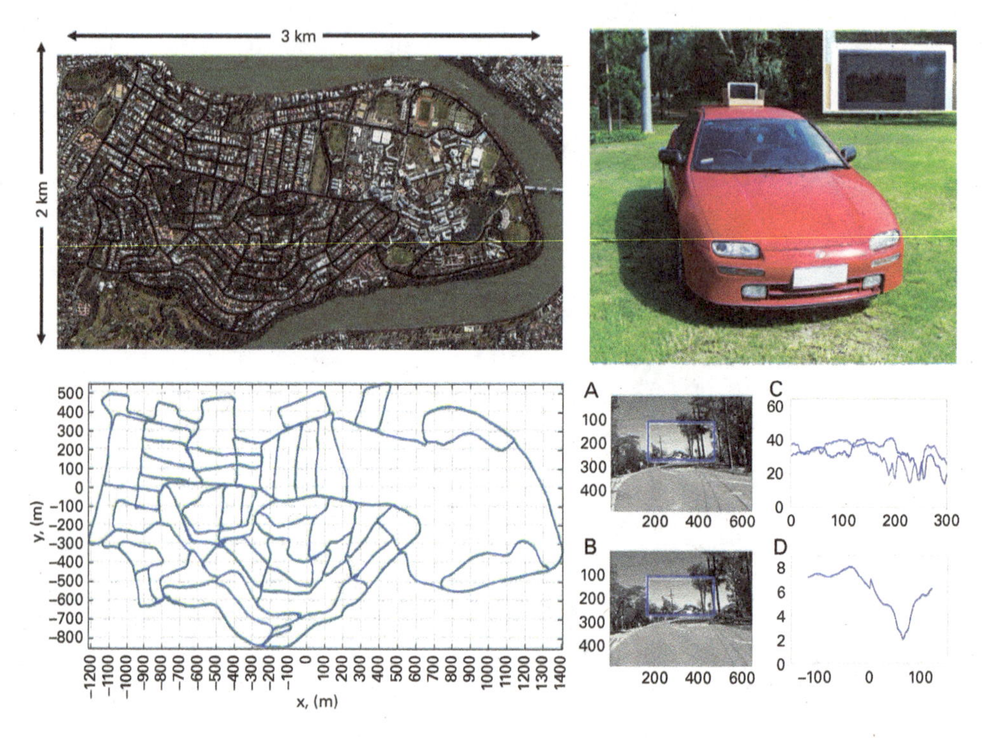

Figure 8.21
Test of RatSLAM in an outdoor suburb. *Top left*: Satellite view of area covered. *Top right*: Laptop with a webcam was mounted on top of a car that navigated the road. *Bottom left*: Resultant RatSLAM map after training. *Top right*: Car used in experiment, with mounted laptop running RatSLAM. *Bottom right*: Place recognition captured from cropped and processed camera images and visual odometry capturing car rotation and translation. Adapted from Milford and Wyeth (2008).

weather, and in any season. The RatSLAM philosophy maintained a balance between faithful representation of the modeled biological systems and state-of-the-art performance. Initial stages of RatSLAM started with neural network models of the mapping processes observed in the rodent brain. Pragmatic modifications were made to support larger and more challenging environments, which seemed to move the model further away from biology. For example, the decision to use pose cells was made because the neuron types known at that time, such as place cells and head direction cells, were unable to represent and correctly update multiple robot location hypotheses. Subsequently, neuroscientists discovered grid cells (Derdikman & Moser, 2010), which share similar characteristics with pose cells. Had the RatSLAM team abandoned the biological neural network completely and moved to a conventional technique, it might have been harder to make the grid cell prediction. However, if they had maintained a more biologically faithful model, they may never have been able to test it in environments that were sufficiently challenging. By following the middle ground, they were able to make contributions in both fields.

8.7 Summary and Conclusions

Navigation is important for survival, and thus many brain areas are devoted to finding one's way around. During navigation, robots face similar challenges to those faced by animals. However, whereas conventional robotics use algorithms to solve navigation tasks, animals need to use some form of neural computation to navigate. Neuroscientists have found a variety of neural correlates of space that include place cells, grid cells, head direction cells and others. These representations have inspired many neurorobotic navigation applications.

It should be noted that although navigation has practical applications, many navigation algorithms can generalize to problem solving. The notion of a cognitive map can be extended from navigating space to navigating semantic information, puzzles, and other mental challenges.

9 Developmental and Social Robotics

9.1 Introduction

As we have seen, studies in neurorobotics often reveal unexpected insights when cognitive models are tested in situated environments. These insights occur more often when the environment is enriched by other agents, whether these other agents are robots or people. This social interaction can lead to complex dynamics that could not have been predicted prior to testing. Modeling social cognition in neurorobots is particularly interesting because their testing environments may include agents that the robots themselves are trying to model. For instance, a neurorobotic model for recognizing and expressing human emotion may be best tested by interacting with actual humans. In such an environment, the robots learn by observing the humans and simultaneously test the validity of their models by monitoring the humans' responses. Modeling social interactions with neurorobotics not only helps us understand social cognition but also helps in developing new ways of assisting humans in everyday tasks. In being able to parse the communication of others and to communicate with others in return, robots can cooperate or compete more effectively. The method of communication may vary in sophistication, from physical gestures and facial expressions to spoken and written language. Furthermore, as agents learn and develop, their communication methods can grow more sophisticated. In this chapter we cover neurorobotic approaches to social development and interaction, including affective robotics, imitation learning, and language.

9.2 The Psychology and Neuroscience of Development and Social Cognition

Cognitive development has long held the interest of psychologists, with several theories developed from behavioral studies. One of the most foundational theories, Piaget's theory of human development, occurs in four stages (Piaget, 1971):

1. Sensorimotor. The first two years of life consists of learning how to perceive and interact with the world. This consists of building reflexes, coordinating complex movements, and building up schemas through constant interactions.

2. Preoperational. From ages two through seven, symbolic representations are formed. This involves object manipulation and speech. Most cognition is egocentric, with limited abilities in theory of mind, that is, being able take the perspective of others.

3. Concrete operational. From ages of seven to eleven, skills of logic and classification develop. Cognition extends beyond egocentric, with the ability to frame things from others' points of view.

4. Formal operation. After age eleven, humans continue to develop their capabilities in abstract thought, metacognition, and advanced problem solving.

The stages show how higher order cognition rests on prerequisite functions. An agent must first learn to process the world on its own terms and then to represent it. Only then is the agent capable of extrapolating these states of cognition to other agents and using representations for communication.

The same framework of development applies to emotional and social cognition and can be used to make sense of neuroscience findings on brain areas and neural mechanisms responsible for these functions. For instance, to understand the emotions of others, an agent may start by feeling those emotions itself.

The feeling of emotion depends on a combination of multiple mechanisms. The **amygdala**, which is part of the brain's limbic system, associates emotions such as fear with memories during consolidation. The amygdala has strong connections with the prefrontal cortex and ventral striatum for functions tying together decision making, motivation, and emotional processing (Bechara et al., 1999; Cardinal et al., 2002; LeDoux, 2000). As discussed in previous chapters, the neuromodulatory system also plays a large role in altering cognitive states such as reward, risk, attention, and arousal. Interestingly, the amygdala is strongly interconnected with neuromodulatory systems.

Extending beyond an agent's individual reactions, the next stage of development involves learning how to parse the mental states of others. **Mirror neurons** in the premotor cortex of primates are observed to fire not only when animals are performing actions such as grasping and reaching but also when observing other animals performing the same actions (Rizzolatti & Craighero, 2004). Figure 9.1 shows an instance of a mirror neuron activity in response to both observing and performing grasping movements. The presence of such neurons supports the idea that models of others' behaviors and states of mind can be represented neurally and be used in planning and decision making. Original studies of mirror neurons focused mostly on motor processing. Others have attempted to tie the mirror neuron system to imitation learning and language acquisition.

However, the mirror neuron system has been challenged when it comes to action understanding (Hickok, 2009; Hickok & Hauser, 2010). It may be better explained by strong sensorimotor integration and the existence of internal models in the brain. Observed actions can serve as important inputs to action selection, including but not necessarily limited to mirroring actions. Sensorimotor integration may be a better and simpler interpretation of

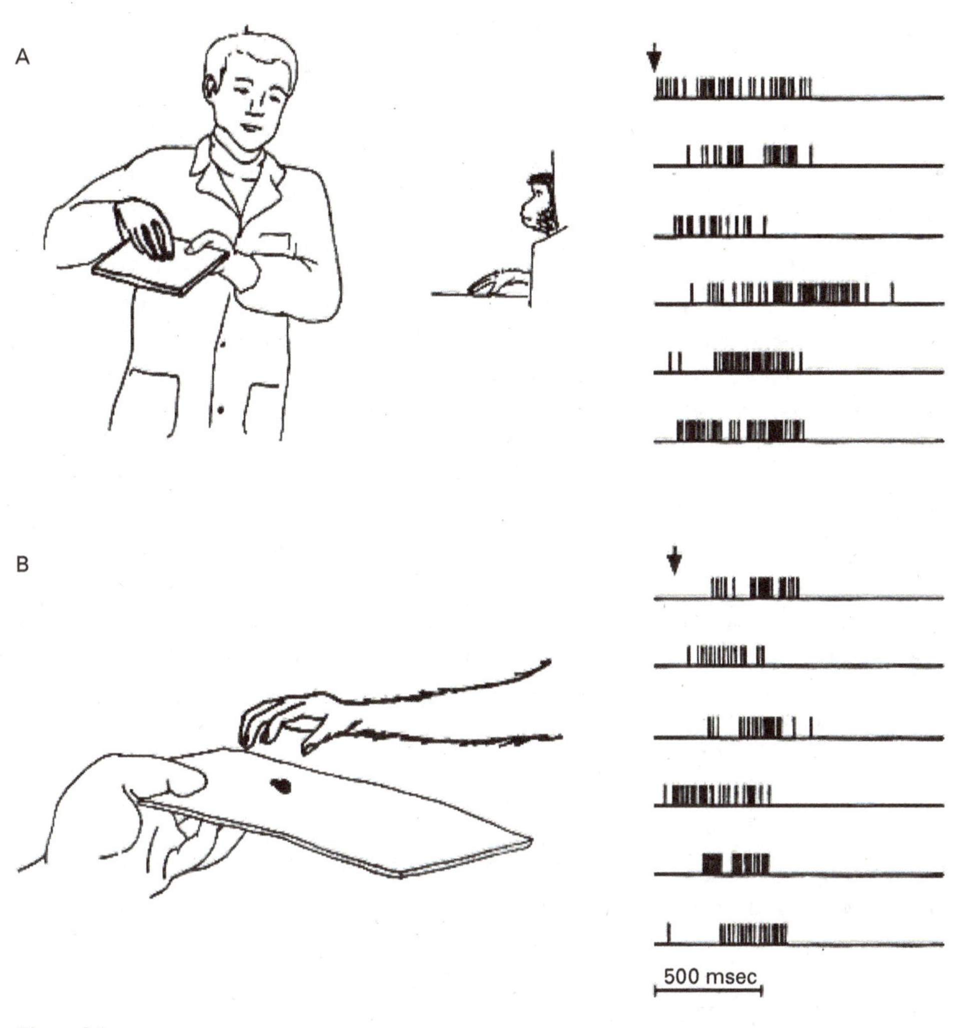

Figure 9.1
Mirror neurons in the premotor cortex of a monkey fire in response to movements by both the self and others. *A*, Instances of a mirror neuron firing in response to observing an experimenter performing grasping movements. *B*, Instances of the same mirror neuron firing in response to performing the same grasping movements. Adapted from http://www.scholarpedia.org/article/Mirror_neurons.

how actions can be understood and imitated. Such integration might be extended to emotional content. In an fMRI experiment, the anterior insula was active for subjects both in directly experiencing an emotion and in observing the facial emotions of others, perhaps serving as a basis for human empathy (Carr et al., 2003).

Understanding the mental states of others enables an agent to communicate with them. Communication can be performed through gestures, speech, or a combination of modalities.

Challenges in communication involve knowing how much information to communicate and in what manner. To know how much information to communicate, it is necessary to develop a **theory of mind**, or a model of the beliefs, desires, and intents of another agent. Because theory of mind is a multidimensional task, neural mechanisms are difficult to pin down. However, areas such as the anterior cingulate cortex, left temporoparietal cortex, and right temporoparietal junction have shown different patterns of activation in response to thinking from the self-perspective or theory of mind perspective (Vogeley et al., 2001). In order to communicate the actual information, there must be a way to map perceptual input to common representations of information. Many believe that the human ability of pointing at objects enabled the evolution and development of language by allowing multiple humans to refer to the same things (Tomasello et al., 2007).

Existing work in development and social cognition in psychology and neuroscience provides an excellent starting point for explorations in neurorobotics. However, a weakness of these existing studies of cognitive development is the inability to make controlled interventions and study their effects. Such interventions would be impractical and possibly unethical in many cases. Fortunately, neurorobotics allows us to make such interventions on simulated agents operating in realistic settings. Although not every aspect of human interaction can be replicated in robots, we are still able to make progress through testing on humanoid robotic platforms, allowing us to observe the subtleties of human gesture, emotion, and object manipulation. In the following sections, we will look at neurorobotics studies involving affect, imitation learning, and language.

9.3 Affective Robotics

Certain aspects of cognition seem uniquely human and are particularly difficult for robots and artificial intelligence to emulate and explain. A primary example is affect, more commonly known as emotion. As discussed earlier, the amygdala and its connections to other brain areas forms an important circuit for emotional processing and memory (LeDoux, 2000). Emotions are states of mind, which are important for social expression and self-awareness. From the self-awareness perspective, emotions can reinforce the stimulus preferences of an agent, encouraging or discouraging certain behaviors. From a communication perspective, the expression of emotions allows agents to estimate each other's motivations through observation and response. Models of neurorobotics aim to capture the cognitive mechanisms behind emotion and the expression of it.

Emotions come in many dimensions and intensities. The intensity of an emotion is linked to levels of arousal, involving the neuromodulatory circuits in the brain. Balkenius and Johansson note that a robot able to change its levels of arousal can react to changes in the environment and modulate decision-making behaviors such as exploring and exploiting (Balkenius et al., 2019). In addition, robots that detect arousal in other agents are able to modulate their interactions accordingly. In Johansson et al. (2020), their robot Epi displays arousal by adjusting its pupil size via a unique eye design of overlapping blades as irises (figure 9.2). The pupil size of the robot reflects its mental state of alertness and arousal while performing tasks of decision making and problem solving. Brain regions supporting a model pupil dilation, including the amygdala and the neuromodulatory area known as the locus coeruleus, which is the source of norepinephrine, are also brain regions important for emotion (Johansson and Balkenius, 2018).

Beyond eye movements, the addition of other facial actuators can enhance a robot's expressiveness, displaying a larger range of emotions from happiness to sadness. In one experiment, the robot's cognitive model learned how to copy human expressions by emulating their facial movements (Boucenna, Gaussier, and Hafemeister, 2014). By closely controlling the physical expression of emotions in response to social stimuli, the understanding between humans and robots can be made more instinctual, leading to more trust and usability of automated systems. This is supported by the suggestion that humans tend to favor interactions with robots displaying positive emotions over negative emotions (Kirby et al., 2010). These studies in affective robotics demonstrate how including emotional

Figure 9.2
Epi is a humanoid robot developed by LUCS Robotics Group at Lund University in Sweden (https://www.lucs.lu.se/research/lucs-robotics-group/epi/). It is designed to be used in developmental robotics experiments. The irises of its eyes can change color and the pupils can dilate and contract.

responses can improve human-robot interactions and the explainability of underlying cognitive models.

9.4 Imitation Learning

Learning facial expressions by observing human expressions is an example of imitation learning, in which expressions or skills are learned by mimicking the actions of others. Complex movements that would be difficult to learn through trial and error can simply be passed along through observation and imitation. Imitation learning is an important aspect of development in social cognition as well, upholding cultural habits and tendencies. In the brain, imitation learning correlates with the functioning of the mirror neuron system, which has inspired many neurorobotic experiments. For example, Billard and Matarić created a robotic system using a hierarchical neural network, with each level of the hierarchy learning the function of a brain region associated with motor control (Billard & Mataric, 2001). Through this model, the robot was able to observe and replicate human arm movements.

Aside from learning complex movements through observation, mirror neurons are also involved in understanding the intentions of another agent. In a neurorobotic experiment (see figure 9.3), a model of the mirror neuron system used a spiking neural network to perform arm movements (Chersi, 2012). The robot observed gesture commands, and then imitated them in a series of phases. During the observation phase, researchers set up an area consisting of colored graspable objects. Then, the researchers performed a gesture to signal the robot to begin the next phase of learning. Two types of actions were taught by the humans, *eating* and *placing*. During *eating*, the robot would reach for an object, grasp it, and put it into a box attached to its body. During *placing*, the robot would reach for an object, grasp it, and put it down in a specific place. Sequential frames of these actions are shown in figure 9.3. In the succeeding imitation phase, the robot needed to look at the cues in the scene to understand which action to perform and then carry out the action. The robot was successful in completing these phases, helping to test theories of the mirror neuron system in imitation learning. For instance, the findings from these experiments provided support for the chain model hypothesis, which states that motor neurons in the parietal cortex and mirror neurons in the premotor cortex are connected in chains that encode motor sequences to achieve selected goals (Chersi, 2012).

9.5 Language

Although the expression of affect provides a fast and intuitive way to communicate intentions, this is sometimes insufficient for conveying higher order thoughts. For this a system of language is needed. The brain requires sophisticated machinery to enable such a high-level process. During language communication the brain must multitask, organize internal

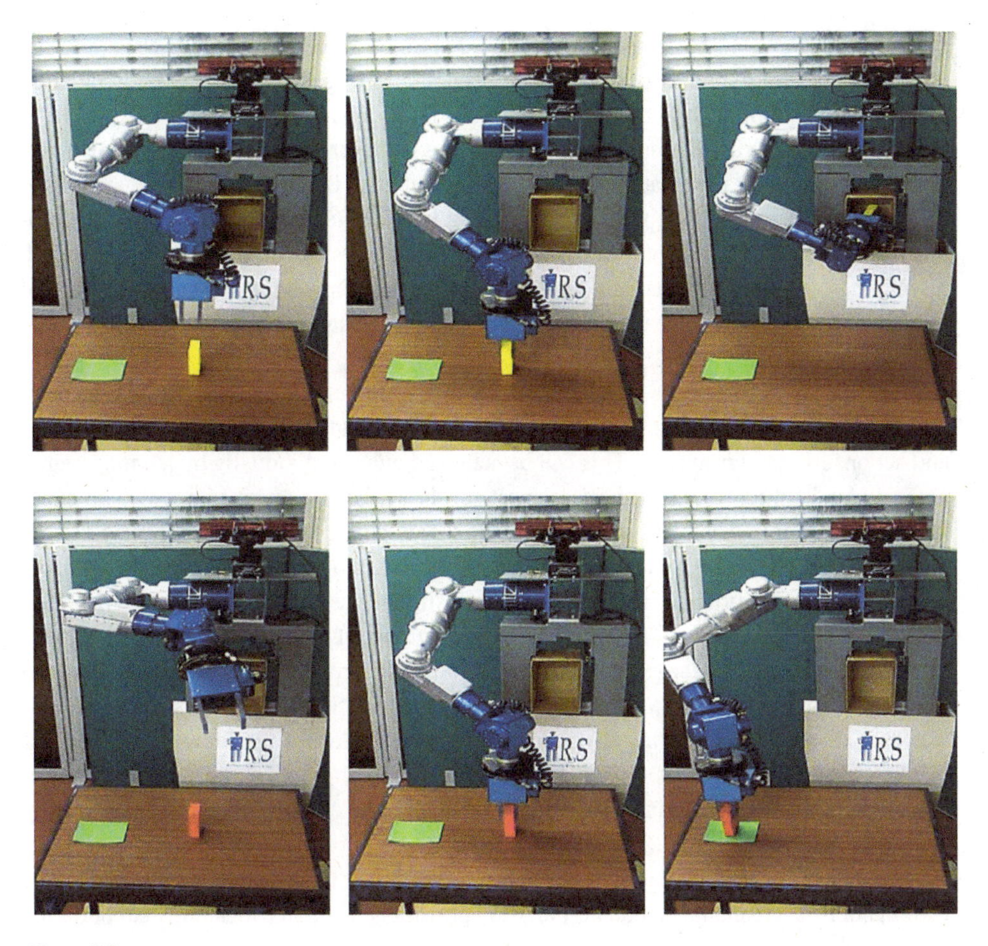

Figure 9.3
Imitation learning experiment with a humanoid robot. *Top*: Successive frames showing the action *grasping to take*, in which the robot picks up the yellow object in order to place it in the robot's own box. *Bottom*: Successive frames showing the action *grasping to place*, in which the robot picks up the red object in order to place it in on the green square. Adapted from Chersi (2012).

thoughts into symbolic sequences of words, utter these words coherently to another agent, and parse the resulting response from that agent. Moreover, the fundamental question of how the external world converts to neural representations is an ongoing line of study. One challenge is that it can be difficult to run controlled experiments to study language acquisition in humans. With neurorobotics experiments we have a means of creating such controlled environments to observe how language emerges, as the robot model can start learning without any prior bias regarding syntactic and symbolic representations. Moreover, compared with pure simulation studies, neurorobots can learn language in an embodied manner, connecting the external world to sensorimotor processing.

How humans were able to evolve speech capabilities in the first place is an interesting question. Oudeyer explored how a system of phonetics might emerge using a population of interactive babbling robots (Oudeyer, 2006). Each robot had a vocal tract capable of uttering phonemes via different neural activations and an ear module capable of converting auditory input to neural activations within a perceptual map. By combining information from the motor and perceptual map, the robots were able to connect phonetic sounds to the production of those sounds. From the population, a shared system of vocalizations emerged, resembling the vowel systems found in human languages. This experiment showed the vital role social interaction and sensorimotor integration played in developing and using spoken language.

From the building blocks of phonemes, words with semantic meanings are formed. Building words within a language requires the grounding of these words with real objects and actions in an environment. This can be achieved either individually by learning paired associations between words and experiences, or through social means, in which an experienced mediator can explain the meaning of words and guide an agent toward correct understanding. In the second case, a mediator who has more knowledge of the language than the learner can provide interactive feedback. In a series of experiments with the Sony AIBO robot (see figure 9.4*A*), Steels and Kaplan studied the impact of learning language through an expert mediator (Steels & Kaplan, 2000). The robot was able to make associations between object views and words via reinforcement learning because the mediator was providing positive and negative feedback to help make the proper associations. The results of these experiments revealed the effects of social feedback on reinforcement of associations for word learning, which drove the robot to explore the world intelligently and make necessary causal inferences.

Next in the language hierarchy is combining sequences of words to form sentences. The combination of elemental language parts to form more complex meanings supports the

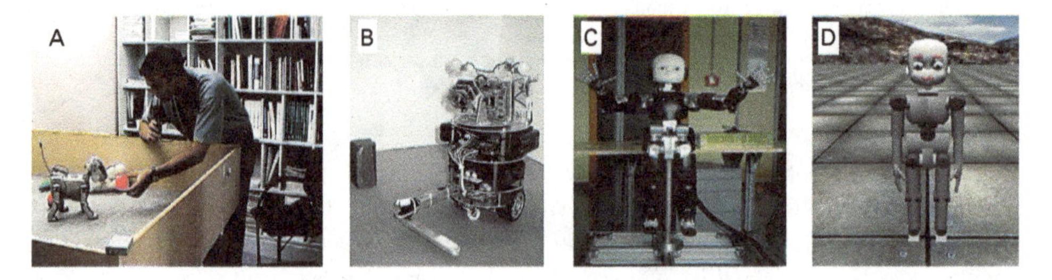

Figure 9.4
Language acquisition and development experiments using robots in embodied environments. Experiments involve the association between motor and cognitive controls. *A*, A human mediator guides the Sony AIBO robot in language tasks to teach it new words (Steels and Kaplan, 2000). *B*, A mobile robot from Sugita and Tani performing actions such as pointing, pushing, and hitting according to language-based instructions (Sugita and Tani, 2005). *C–D*, The iCub robot, real and simulated, acquiring language multimodally. Adapted from Tikhanoff et al. (2011).

idea that language is largely compositional. To study compositionality, Sugita and Tani (Sugita & Tani, 2005) used a connectionist model on a mobile robot to show how linguistic and behavioral processes can be combined to learn higher-level meaning (see figure 9.4*B*). In their model a recurrent neural network (RNN) was used to classify and create novel sequences of words and motor actions. In the training phase the network connected sentences and their associated behavioral sequences. In the testing phase the robot was asked to perform behaviors according to sentence-based commands. The robot showed good generalization of the training instances, correctly responding to commands not previously encountered. This was possible only through compositionality, which is the novel combination of previously learned language components. More specifically, the robot was able to make appropriate distinctions between nouns and verbs, correctly assembling the meanings of the words to understand full sentences. The compositional capabilities were achieved through the combination of linguistic and behavioral components, leading to the self-organization of dynamic linguistic structures.

The purpose of language is often to understand instructions or provide instructions to achieve goals. To demonstrate how neural networks could integrate speech and action, the iCub robot platform was used to learn names for actions and objects (Tikhanoff et al., 2011). In their experiments, a feedforward neural network used backpropagation to learn how to reach toward objects, and a recurrent neural network was used to learn how to grasp the objects (figure 9.4*C–D*). Inputs from the robot's visual system were segmented to find objects, and a real-time speech recognition program processed verbal commands. Another neural network was used for goal selection, deciding between the actions for idle, reaching, grasping, and dropping. The robot learned the correct association between verbal instructions and action sequences, showing the involvement of multiple modalities of sensing and acting in language learning.

9.6 Social Robotics: Applications and Outreach

Because social robotics involves interactions with people, it naturally extends to a wide variety of applications benefiting people, including therapy for developmental disorders and science, technology, engineering, arts, and mathematics (STEAM) education. The humanlike functioning of neurorobotics makes them applicable to therapeutic treatments related to social activity. For example, the caretaker robot (CARBO) from Krichmar and Chou was explored as a means of providing sensory integration therapy for developmental disorders such as autism spectrum disorder and attention deficit hyperactivity disorder (Krichmar & Chou, 2018). The robot was able to perceive tactile input through an array of trackball sensors on its body (see figure 9.5). In a series of games that taught users how to interact with the robot, associations between touch, emotional response, and social interaction were trained. In general, socially assistive robots serve as a good tool for users to learn interaction

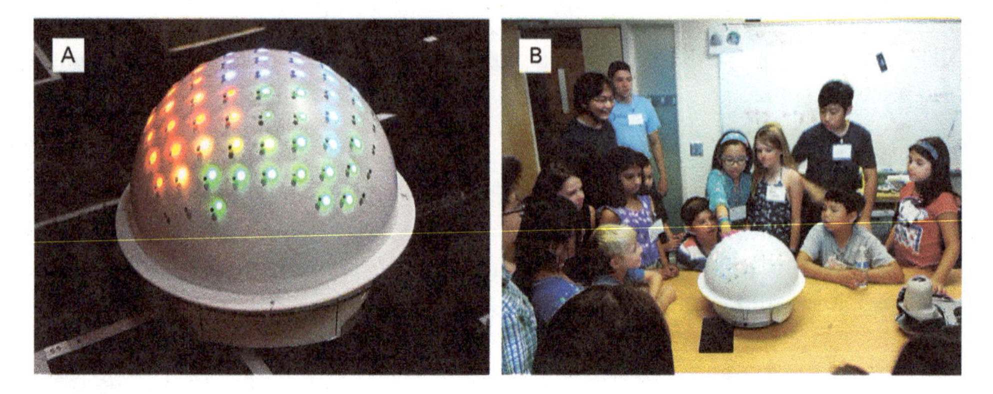

Figure 9.5
CARBO the caretaker robot. *A*, The shell had an array of tactile sensors that consisted of trackballs and colored LEDs. *B*, Children could rub or pet the robot's shell and in response it would light up with different colors.

skills with less social pressure and intimidation than interacting with an actual human (Mataric & Scassellati, 2016; Tapus et al., 2007).

Interestingly, how well a person interacts with a robot may depend on the person's bias. In a human-robot interaction study in which people had to assess whether the robot had intentions, brain activity prior to the interaction could predict the person's bias toward the robot (Bossi et al., 2020). Subjects were shown pictures of an iCub robot and asked whether the robot's behavior could be attributed to its mechanical design or having some intentions (see figure 9.6). **Electroencephalogram** (EEG) activity prior to the task could accurately predict a subject's attitude toward the robot.

Neurorobotics not only helps in therapeutic applications but also benefits the public at large through educational demonstrations (see figures 9.5*B* and 9.7). Neurorobots explain the inner workings of AI to people who may have had little exposure to computer science and neuroscience beyond popular media depictions. Neurorobots also provide an interactive means for researchers to promote their field of study. As AI becomes more prevalent, it requires trust from those who use the technology. Interaction with robots adds an intuitive explainability to the technology (Chen et al., 2020). In contrast to more conventional methods of explainable AI from the field of machine learning, neurorobots interact with humans in the real world. Additionally, these demonstrations can often be produced from scientific experiments that have already been completed. For example, the CARL robot from the Cognitive Anteater Robotics Laboratory (see figure 9.7*A*), which was used in several research experiments, has also been used in numerous public demonstrations for visitors, explaining concepts of neuromodulation to a broad range of audiences (Cox & Krichmar, 2009). Concepts and algorithms that appear complex can be easier to explain when the robot is physically performing humanlike tasks. As CARL was observed to move towards green stimuli and away from red

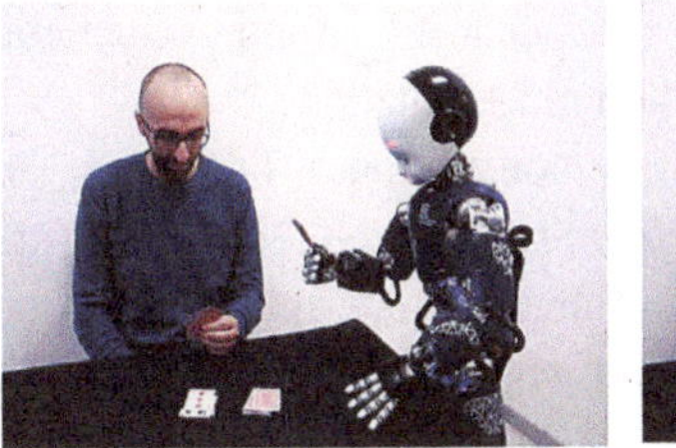

Figure 9.6
Attitude bias to robots. Human subjects were presented with scenarios like the one depicted in the figure. They used the slider to decide whether the scenario was due to the mechanical design of the robot or the robot's intention. EEG was used to predict the subject's bias. Redrawn from Bossi et al. (2020).

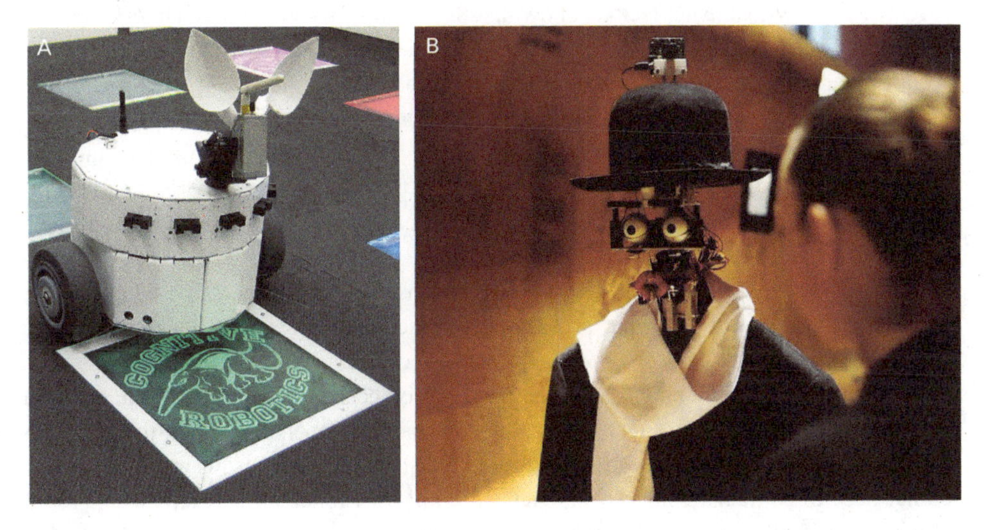

Figure 9.7
Examples of robots used in public outreach demonstrations. *A*, CARL interacting with its colorful floor was demonstrated to adults and children many times to explain how neuromodulation in the brain can lead to attention. *B*, Berenson tours an art museum while expressing emotions through facial expressions, reacting to stimuli such as artwork and other people. Adapted from https://news.artnet.com/art-world/robot-art-critic-berenson-436739.

stimuli, visitors attributed the behavior to human-like emotions such as fear and excitement, fostering an interest in learning more about the details of the implementation.

The robot Berenson (figure 9.7*B*) was used in a public demonstration at the Musée du Quai Branly during an exhibition titled *Persona: Oddly Human+Emotion* (Moualla et al., 2018). Using its facial actuators, the robot expressed its opinion of the art works at the museum, bringing about interesting questions about how and whether AI could really perceive and form an opinion about art.

9.7 Case Study: Emotional Interactions

Robots that can express emotions through facial actuators are not only able to communicate more effectively with other social agents but also play a major role in understanding learning and development. Positive and negative emotions can be used to socially transmit value judgments of environmental factors, allowing other agents to learn with much less trial and error than when learning alone. For human babies facial expressions from their caregivers are a primary source of feedback. From the neurorobotics perspective it is an interesting topic to explore on an embodied platform, because it involves constant back-and-forth interactions, regarding aspects of a physical environment, between agents. In this case study we examine Boucenna and colleagues' insights into how facial expressions contribute to cognitive development and preference learning (Boucenna, Gaussier, et al., 2014; Boucenna. Gaussier, and Hafemeister, 2014).

The experiment consisted of a robotic agent with the ability to perceive human facial expressions, make its own facial expressions, and reach for objects. A human experimenter was tasked with teaching the robot to grasp desirable objects and avoid undesirable objects. To achieve this, the robot had to learn the values of objects by understanding the emotion of the human's expression. Figure 9.8 shows the basic components of the robot's brain model. Some components of the model are reflexive, that is, are preexisting associations that do not need to be trained. Other components are trained via human interaction. The robot starts with the ability to make facial expressions according to its internal states of pain and pleasure. It also starts with the knowledge that positive objects should be grasped and negative objects should be avoided. When the robot experiences emotional states, it forms the corresponding facial expression. The human then mimics the facial expression while the robot observes. This is akin to caregivers empathetically making the same expressions as the babies. After some time, the robot begins to associate the human's facial expressions with the corresponding positive and negative internal states. Additionally, these states are associated with the objects presented to the robot. That is, if a human has a positive expression while an object is in view of the robot, the robot learns that the object is positive and should be grasped, and vice versa with negative objects. For example, if the human caretaker smiled while presenting a water bottle, the robot learned to smile and grasp the water

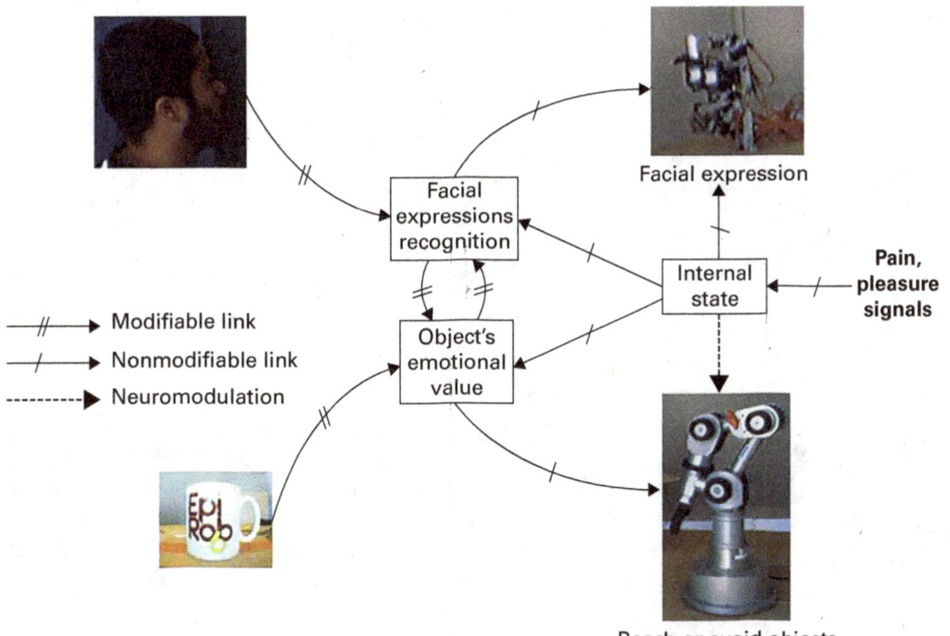

Figure 9.8
Overview of network model for learning to grasp desirable objects. The robot learns facial expressions by making facial expressions that match their internal state and observing the human who is mimicking the robot's expression. Then, when viewing an object and a human's facial expression together, the robot learns whether the robot has positive or negative valence, which influences the robot to either reach toward or avoid the object. Adapted from Boucenna, Gaussier, and Hafemeister (2014).

bottle when it was presented. However, if the caretaker grimaced while holding a toy gun, the robot frown and move its arm away from the gun.

Figure 9.9 shows examples of facial expressions made by the robot and the robot imitating a person's expression. The four expressions happiness, sadness, surprise, and anger were used in the experiment. The robot parsed facial expressions by finding the difference of Gaussians (DoG) features in the images, extracting points of visual interest. The presence of these features was associated with the emotions experienced by the robot. Initially, the robot experienced these emotions at random, and the human attempted to imitate those emotional expressions. After multiple trials of this the robot was able to cluster the visual features into the four emotion categories, thereby learning to recognize the expressions in human faces. The model was robust enough to recognize these emotions in multiple humans.

Figure 9.10 shows details of the neural network used in the experiment. As in the development of a human child, the robot learned in a succession of phases, outlined by boxes in the figure. The robot started by learning the task of facial expression recognition. Once

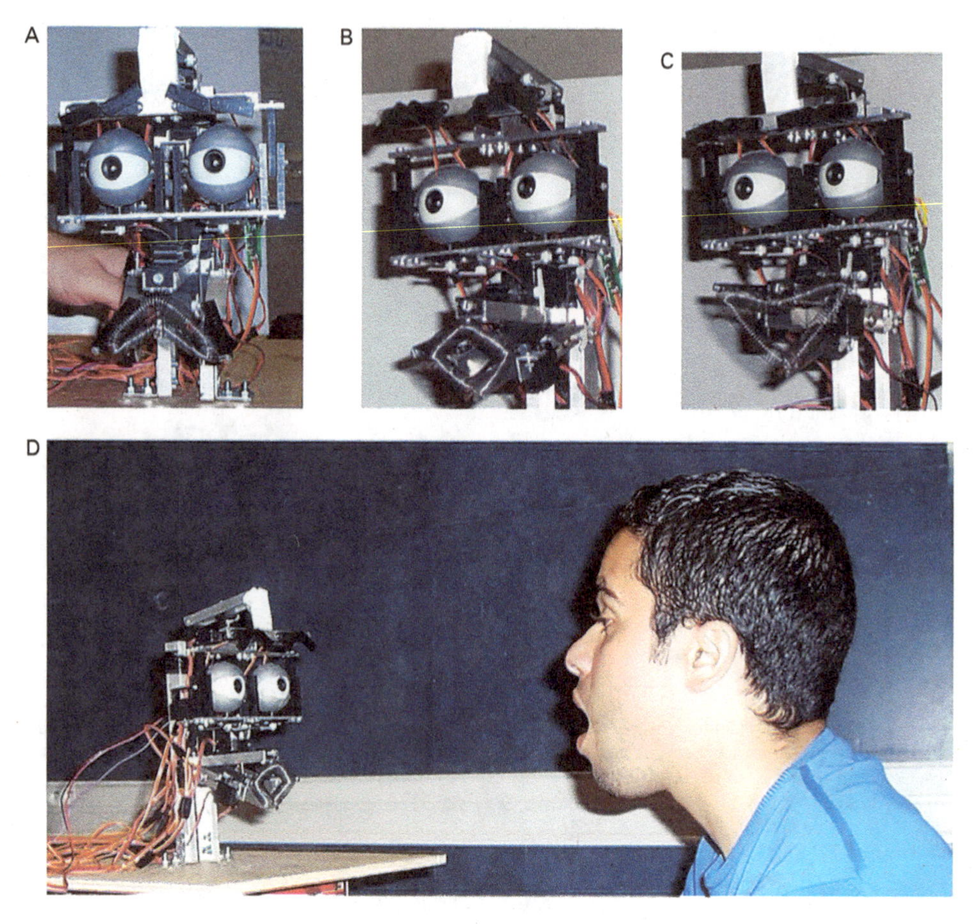

Figure 9.9
A, Sad expression. *B*, Surprised expression. *C*, Happy expression. *D*, Robot learning facial expression via imitation learning. Adapted from Boucenna, Gaussier, and Hafemeister (2014).

expressions had been learned, they were then associated with the objects. The object valences were then used as a modulator for learning grasping movements, such that a positive valence increased the amount of attention paid to the object. The arm was then encouraged to be centered toward the object while grasping. The opposite was true for negative objects, such that the robot's attention and arm moved away from the object. This experiment showed how imitation learning of facial expressions may serve as a key learning mechanism, allowing agents to learn skills by gathering valence information from social cues. These skills are complex and require a multistage development process, starting from the early stage of learning how to parse emotions from facial expressions to the later stage of associating these emotions with physical stimuli. One could imagine how the development

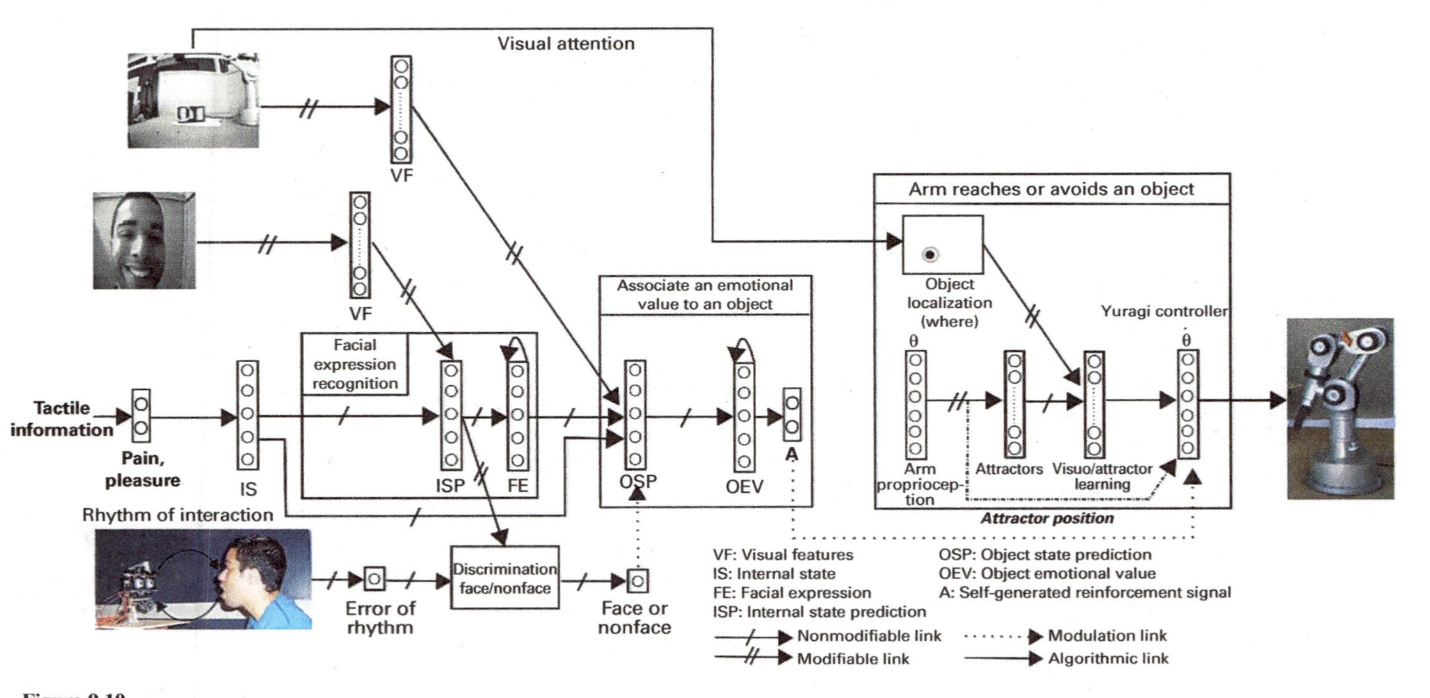

Figure 9.10
Complete network architecture of model. Three stages of learning occur: (1) the recognition of emotions from inputs of human facial expressions, (2) the association of emotional values to presented objects, and (3) the reaching toward or avoidance of objects with positive or negative valence. Adapted from Boucenna, Gaussier, and Hafemeister (2014).

process could expand to even more complex skills of object manipulation, problem solving, and planning.

9.8 Case Study: Grounding Actions to Words

Language is another means of social communication, supporting the capability to express complex concepts and sequences. Because language relies on the combination of symbols such as phonemes or visual characters to communicate, the question arises of how these abstract symbols are grounded to represent concepts in the real world. Specifically, for actions, which are not tied to singular objects, the problem is dynamic, involving the interactions of multiple objects over time. In this case study we examine the work of Marocco and colleagues in which the iCub humanoid robot interacts with objects and observes the outcomes, tying the actions and observations into linguistic labels (Marocco et al., 2010).

The iCub, a bipedal child-sized robot, was used in the study. Much of the work involving the iCub has focused on the study of cognitive development, including language. A simulated version of the iCub was used in this experiment, in which the robot interacted with three objects by pushing them and observing the outcomes. The three objects consisted of a sphere that rolled when pushed, a cube that slid when pushed, and a cylinder that was fixed to a flat surface. The cognitive model consisted of a neural network encoding sensory-motor sequences and linguistic labels, as depicted in figure 9.11. The network contained eight input units, ten hidden units, and eight output units. The output represented the input at the next time step, capturing sequences. Of the eight input and output units, three units encoded joint information of the shoulder and neck, one unit represented tactile information that is output if the hand is touching an object, one unit represented the roundness of an object, and three linguistic inputs trained as one-hot encoding vectors for each of the three types of object interactions ([1,0,0] for sphere, [0,1,0] for cube, and [0,0,1] for cylinder).

Figure 9.12 shows two example training sequences. In one sequence the robot pushes the sphere and observes that the object continues to roll, and in the other sequence the cube does not roll after being pushed. As the head of the robot is programmed to track the object, the sequence of movements from the head and neck actuators is observed to be in a different pattern according to the object being manipulated.

Figure 9.13 shows sensor activations over time during interactions with the objects. When interacting with the sphere, the actuators show greater dynamism in its head tilt sensor when interacting with the sphere. The touch sensor activation is also very different for the sphere as the ball continues to roll when pushed and loses contact with the robot's hand. These differences were learned by the neural network, resulting in each of the semantic labels being properly associated with the dynamics of each object interaction.

The next question was whether the semantic labels were being associated with the objects or with the dynamics of the actions and movements. Three novel object interactions were tested, as shown in figure 9.14, including a cylinder placed on the round surface such

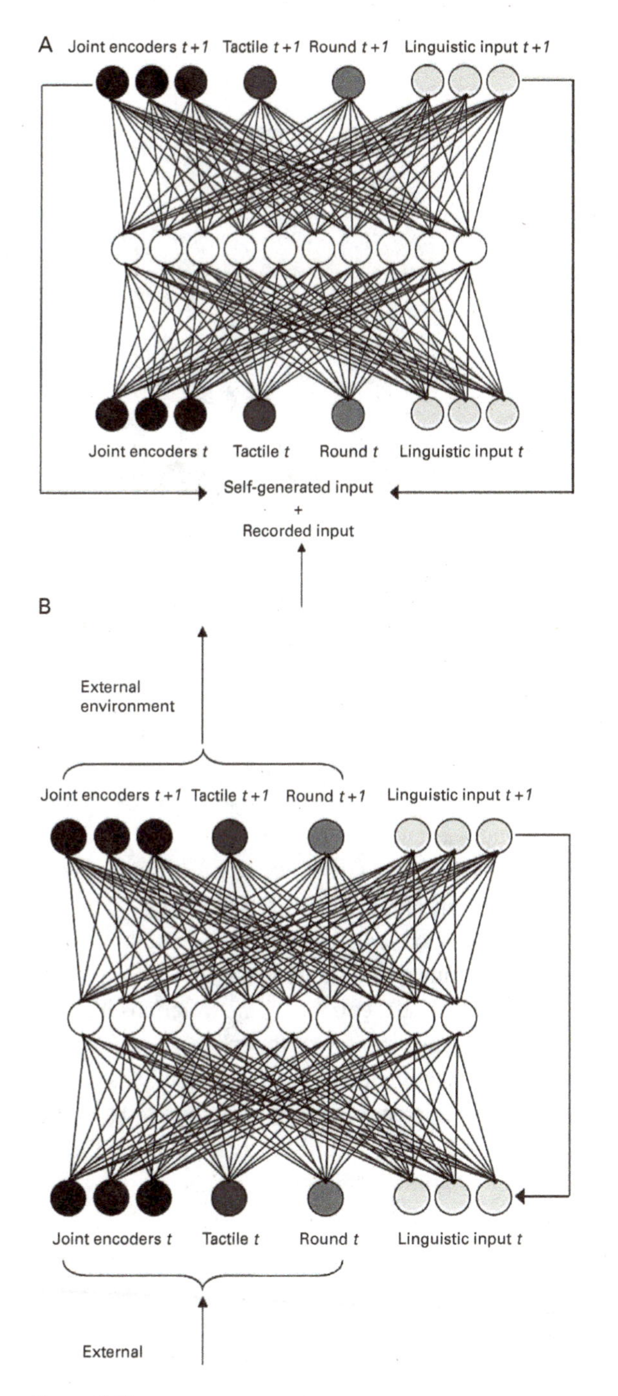

Figure 9.11
Neural network architecture for grounding actions. The input and output both consist of eight units: three units for joint encoders, one unit for tactile sensing, one unit for the roundness of an object, and three inputs for linguistic information. The input represents the current time step, and the output represents the next time step. *A*, Network for training. The output feeds back to the input as the robot learns transitions between its own internal states from recorded data. *B*, Network for testing. Input comes from external stimuli, and output acts upon the external world. Adapted from Marocco et al. (2010).

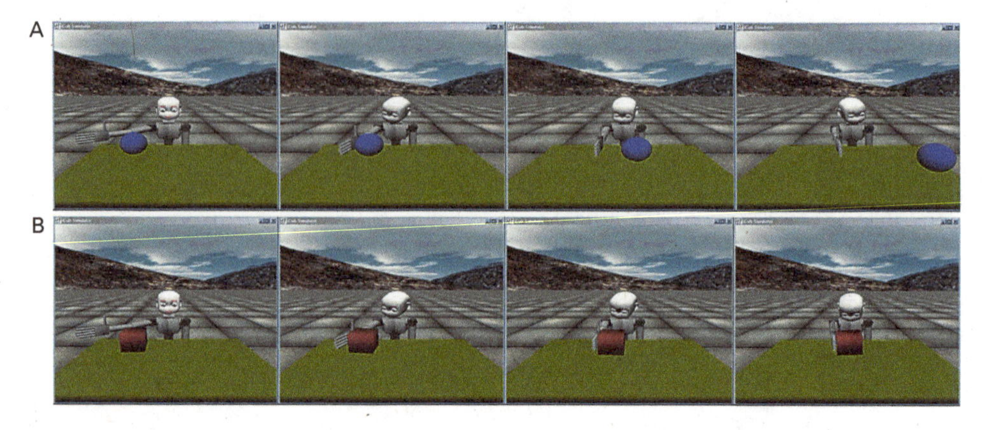

Figure 9.12
Two sequences of the simulated iCub robot interacting with an object. *A*, The robot pushes a blue sphere to the left and observes its rolling movement. *B*, The robot pushes a red cylinder to the left and observes its lack of rolling. Adapted from Marocco et al. (2010).

that it rolled when pushed, a cylinder placed such that it would slide when pushed, and a fixed cube that did not move. For the rolling cylinder the linguistic output associated with the sphere was the most active. For the sliding cylinder the linguistic output associated with the cube was the most active. Finally, for the fixed cube the original cylinder output was the most active. The activities of the linguistic outputs did not settle until the robot had contacted and pushed the object. These results show that it was not the objects themselves that were encoded but rather the actions and the observed results.

This has interesting implications about how linguistic labels are learned and how these labels are associated with our actions. The learning of these semantic labels shows that an embodied environment is necessary to learn the words corresponding to actions because the robot must interact with objects to learn the consequences of the interactions. There is also a distinct difference between learning the meanings of nouns and verbs. Just as the iCub had to interact with the objects to ground the meanings, a human child must do the same as part of the processes of cognition development. It further strengthens the importance of considering embodiment in studying cognitive functions.

9.9 Summary and Conclusions

Development and social interaction are complex processes. In the stages of cognitive development, a child first learns to react appropriately to environmental stimuli, then learns to perceive reactions in others, and finally learns to communicate these concepts via back-and-forth interactions. Throughout the entire process the person or robot is situated in the environment and engaged with other cognitive beings. The subject matter supports many

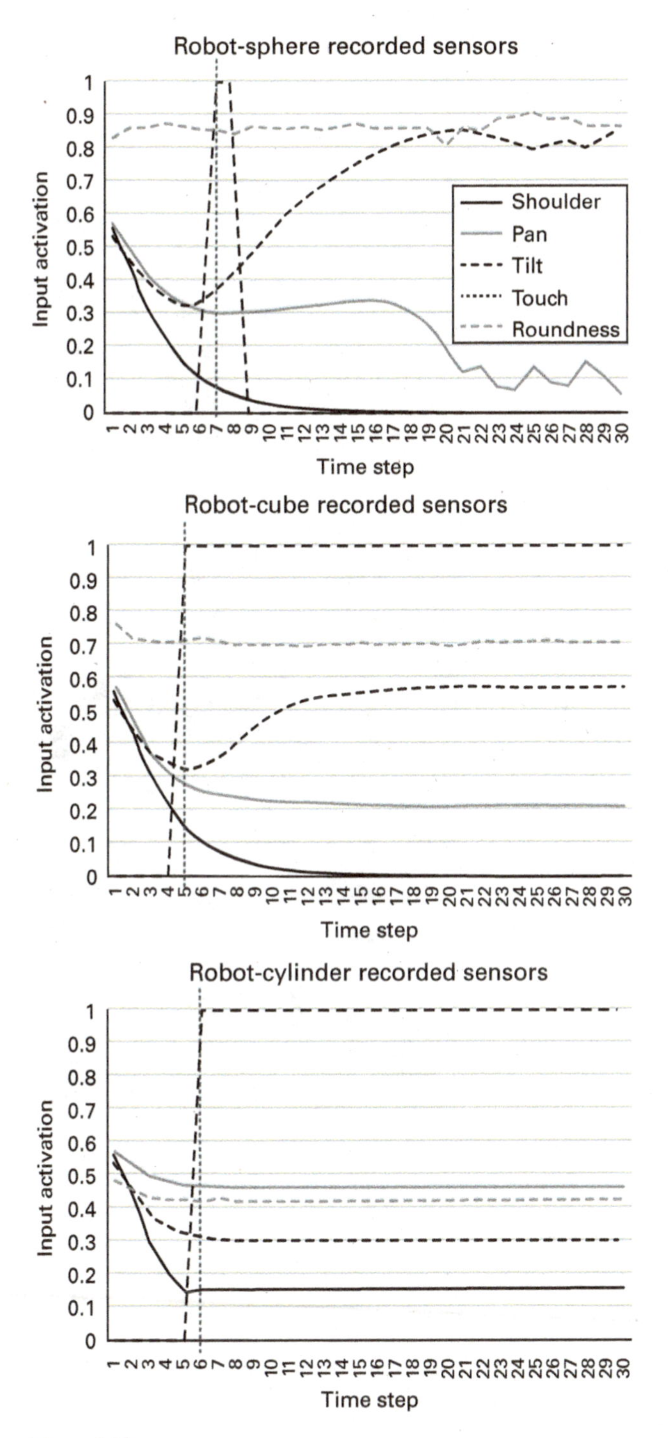

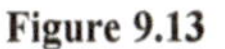

Figure 9.13
Sensor activations of recorded sessions during robot interactions with the sphere, cube, and cylinder. Adapted from Marocco et al. (2010).

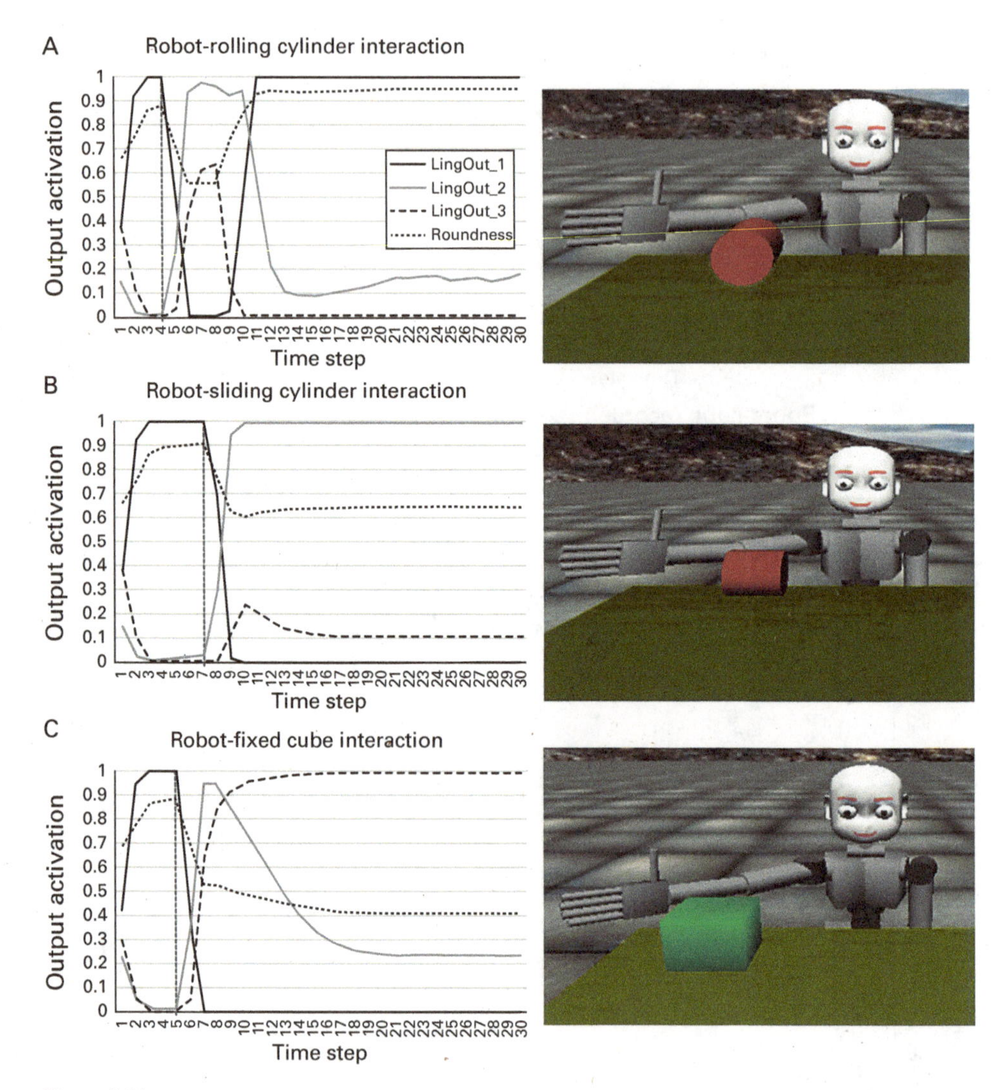

Figure 9.14
Three new conditions used to test the generalization capabilities of the linguistic units. *A*, Interaction with a cylinder placed such that it will roll. *B*, Interaction with a cylinder placed such that it will not roll. *C*, Interaction with a fixed cube. Adapted from Marocco et al. (2010).

opportunities for studies in neurorobotics to reveal new insights on existing theories in neuroscience and psychology.

In this chapter we have seen examples of how facial emotion processing and language emerge in robots from natural interactions with the environment and other agents. These examples highlight the roles of neural mechanisms in representing another agent's cognitive states. Neurorobotics findings in this area are also particularly relevant in developing technologies to forge connections and facilitate communication. Particularly, telepresence technologies for virtual reality and augmented reality explore ways to bring the advantages of physical interactions to virtual interactions. Embodied experiments allow us to map which elements of physical interactions lead to which modes, emotions, and states of understanding. Neurorobotics can help us understand the connections between each other and can ground these connections in cognitive neuroscience.

10 Neurorobotics: Past, Present, and Future

10.1 Introduction

The goals of this book were to introduce readers to the field of neurorobotics and to provide design principles for those who want to conduct research in this field. Neurorobotics is a growing field with great promise. It is an interdisciplinary field that touches on biology, cognitive science, computer science, engineering, neuroscience, psychology, and robotics. Although it may seem that one needs to have extensive knowledge to make progress in neurorobotics, in practice that is not the case. Following the design principles laid out in this book, one can identify interesting directions of research while developing the necessary skills along the way. Furthermore, a single neuroroboticist rarely has expertise in all areas and instead relies on collaborations with experts in other disciplines to develop new ideas. Applying general neurorobotics principles in one's experiments or in one's thinking can contribute to coordinated collaboration across these related subfields.

In our experience, practicing neurorobotics principles can transfer to how one approaches problems in other fields. Neurorobotics is a holistic approach that combines brain, body, and behavior. It forces one to consider how the behavior in an environment may affect the design of the agent, whether it is a robot or a simulation. Because robots rely on local information and their actions take place in an environment, it forces one to consider the limits of sensory input and motor output. Furthermore, when one tests a robot in the real world there is a reduction in potential biases. The real world is complex, dynamic, and noisy, and it does not come with rules. Therefore, it is a more rigorous and realistic test of algorithms.

Neurorobotics is also a powerful tool for testing brain theories and increasing our understanding of neuroscience. The robot controller is modeled after some aspect of the nervous system. Unlike human or other animal studies, the neuroroboticist has access to every aspect of this artificial brain during the lifetime of the agent. Therefore, the neuroroboticist can analyze and perturb the nervous system in ways that a neuroscientist cannot with present recording technology. Not only can a neurorobot be tested under laboratory conditions that are similar to those of an animal experiment in order to provide direct comparisons, but it can also be tested in more natural conditions to see how these brain functions might respond to real-world situations.

It is our hope that interested readers will be inspired to include neurorobotics in their own research endeavors or perhaps use ideas from neurorobotics in how they approach problems and challenges.

10.2 Summary and Takeaways

In this book, we have broadly covered the field of neurorobotics and a wide range of relevant material. Therefore it is worthwhile to review what was presented in all its chapters. Following is a summary of the principles covered in this book:

Chapter 1: Neurorobotics: Origins and Background

The early days of neurorobotics included fundamental experiments such as Grey Walter's tortoises and Braitenberg's vehicles, which produced interesting emergent behaviors similar to biological organisms. These experiments were developed as artificial intelligence and robots began to capture the imagination of the public with science fiction–fueled visions of what robots could achieve for society. At the same time questions in systems neuroscience, such as how the organization and circuitry of the brain contributed to intelligent behavior, caused researchers to view robotics as a means of testing neuroscience theories in an embodied environment. From the questions and curiosities of artificial intelligence and neuroscience, the field of neurorobotics emerged.

Chapter 2: Neuroscience: Background for Creating Neurorobots

Systems neuroscience investigates the circuitry of the brain, from the level of individual neurons to the broader connectivity between brain areas. Neuron models can range from detailed simulations of dendritic branches in a single neuron to more abstract models summarizing the activity levels of neuronal populations. Models of connectivity include network analyses of different connection types and functional neuroanatomy. Brain-inspired neural network models can integrate biological studies of neuron types, brain regions, and anatomical projections within and between brain areas. Models of system level neural architectures show how processes such as learning, memory, pattern recognition, and action selection occur. These models serve as the foundation for neurorobotics experiments.

Chapter 3: Learning and Memory

Learning and memory allow neurorobots to learn from experience and increase their knowledge of the environment for better access to resources and avoidance of danger. Learning comes in categories of supervised learning, unsupervised learning, and reinforcement learning. Popular models of learning in neural network models include Hebbian learning and the Rescorla-Wagner learning rule. Corresponding learning rules exist for spiking neural networks, such as spike timing dependent plasticity (STDP). Neurorobotics experiments produce naturalistic sensory input to these models, providing insights on how associations emerge between sensory stimuli and behavior through interaction with the environment.

Chapter 4: Reinforcement Learning and Prediction

Compared with other types of learning, reinforcement learning is particularly driven by interactions within an environment involving a closed-loop system of states, actions, and feedback in the form or rewards, punishment, or other environmental cues. Rewards come from value systems that maximize the survival of the agent. Models of reinforcement learning can be model based, explicitly learning the transitions between states, or model free, directly learning the optimal actions for each state. As the agent explores its environment and gains knowledge, it increases its ability to predict the future states. Neurorobotics experiments in reinforcement learning utilize real environments and rewards, often yielding interesting and sometimes unexpected results.

Chapter 5: Neurorobot Design Principles 1: Every Action Has a Reaction

In designing neurorobots the actions of the robot are entrenched in the surrounding environment. For instance, simple mechanical designs can perform seemingly complex functions by taking advantage of unique environmental aspects. The designs are also tuned to the environment to have multiple systems capable of performing the same functions. In this way the agent can still survive in the environment should one system fail. Processes run in parallel in an event-driven manner, continuously responding to concurrent events.

Chapter 6: Neurorobot Design Principles 2: Adaptive Behavior, A Change for the Better

To adapt to a changing environment a neurorobot must be able to learn, store and recall information. Due to limitations in learning capacity and memory storage the robot must also be able to decide which pieces of information are important to remember. Memory systems in neurorobotics are particularly applicable in spatial memory for navigation and contextual memory for learning representations of the environment. Navigating within this environment requires the processing of risk, reward, and uncertainty, which is done via neuromodulatory systems in the brain. Through such systems robots are more able to predict future events and adapt to changes in the environment.

Chapter 7: Neurorobot Design Principles 3: Behavioral Trade-Offs Because Life Is Full of Compromises

The world is full of trade-offs and changing needs that require us to make choices. Incorporating behavioral trade-offs such as reward vs. punishment, invigorated vs. withdrawn activity, expected vs. unexpected uncertainty for attention, exploration vs. exploitation of choices, foraging for food vs. defending one's territory, coping with stress vs. keeping calm, and social interaction vs. solitary restraint. All can lead to interesting behavior in neurorobotics. Many of these trade-offs are regulated by neuromodulators and hormone levels.

Chapter 8: Neurorobotic Navigation

Navigation allows robots and biological agents to avoid predators, find food, seek shelter, and adapt to dynamic environments. Traditional robotics contains navigation challenges, which are addressed using algorithms such as simultaneous localization and mapping (SLAM) and path planning. Biological agents, faced with the same challenges, perform the same functions under the constraints of neural circuitry by using spatial representations that include head direction cells, place cells, grid cells and other neural correlates of space. From these spatial representations navigation strategies can emerge through interactions and learning.

Chapter 9: Developmental and Social Robotics

Developmental psychology and human robot interaction can be studied using the neurorobotics approach. In cognitive developmental robotics modeling how children develop perception, movements, and language can provide inspiration for creating better robots. Social robots, which incorporate brain models of emotion or empathy, can make the interaction between humans and machines more natural. This requires methods of communication such as recognizing and performing emotional expressions. Neurorobots capable of communication with humans have the additional benefit of being able to perform community outreach, influencing public perception of robots and introducing the field of neurorobotics to laypeople.

10.3 Neurorobotics Challenges and Contributions

Despite major accomplishments in deep learning and machine learning there are differences between artificial intelligence and natural intelligence that may limit progress. To address these limitations, Jeff Hawkins argued that intelligent systems must incorporate the following key features of the brain (Hawkins, 2017). (1) Learning by rewiring; we learn quickly, incrementally, and over a lifetime. (2) Sparse representations; biological systems are under extreme metabolic constraints and need to represent information efficiently. (3) Embodiment; sensorimotor integration is observed throughout an intelligent system. In addition to these we would add the following (Krichmar, 2018). (4) Value systems; extracting saliency from the environment and responding appropriately. (5) Prediction; using experience to be more successful in the future. These five features have been discussed throughout this book.

The principles of neurorobotics introduced in this book can help address major challenges facing artificial intelligence and robotics research. In general, neurorobotics explores these

Table 10.1
Challenges in AI and contributions of embodiment from neurorobotics

Challenge	State of the art	Limitations	Neurorobotics contribution
Lifelong, continual learning	• Transfer learning • Domain adaptation	• Switching between tasks • Understanding context	• Biologically inspired learning and memory (chapter 3) • Schemas and the interaction between long-term and short-term memory (chapter 6)
Computing power	• Power-hungry algorithms run on GPU clusters and super computers	• Time-consuming training • Power-inefficient deployment	• Bio-inspired efficient computing and morphological computation (chapter 5) • Multitasking and event-driven architectures (chapter 5) • Neurorobotics algorithms compatible with low-power neuromorphic chips (chapter 5)
Knowledge and data scarcity	• Models pretrained on large data sets • Offline training	• Time-consuming training periods • Separation of training and testing • Difficulties transfering knowledge	• Behavioral trade-offs decide what data is needed (chapter 7) • Navigation strategies improve information foraging (chapter 8)
Human-compatible interactions	• Designs based on developmental psychology • Limited deployment outside research labs	• Requires human expertise • Unnatural interactions	• Social affect for natural interactions (chapter 9) • Language grounded in action (chapter 9)

challenges in embodied settings, providing a fresh perspective that extends beyond computations and algorithms performed on machines. Neurorobotics incorporates features of the brain that may begin to address the challenges associated with lifelong continual learning, efficient computing, operating on scarce knowledge, and human-computer interaction (see table 10.1).

10.3.1 Lifelong, Continual Learning

In contrast to natural learning systems, current artificial intelligence applications are typically designed and trained on very specific tasks. This makes adapting to a dynamic environment difficult. For example, robots reliably perform large volumes of repetitive work in factories but would be unable to perform the task if the assembly line changed due to new parts or building instructions. An agent with **lifelong learning** and **continual learning** would be able to learn and switch between multiple tasks, with the ability to adapt and apply prior knowledge to new settings. Moreover, this adaptation would be online, occurring seamlessly during operation without periods of offline training.

Typical machine learning approaches to lifelong learning include transfer learning and domain adaptation, which explore how to enable systems to learn without catastrophic forgetting of prior information. As covered in chapter 3, biologically inspired mechanisms of learning and memory have much to offer in the way of flexible representations and learning. As in the schema experiments described in chapter 6, a model of long- and short-term memory in the hippocampus and neocortex can allow neurorobots to acquire general knowledge from singular experiences in the real world if it fits within a context or schema. However, new schemas are formed when faced with novel and unfamiliar stimuli. This brain strategy may be beneficial in dynamic environments. In general, neuroscience and cognitive science have demonstrated how humans and animals generalize and transfer information rapidly and fluidly. Modeling their brains and behavior may be a key step toward lifelong, continual learning.

10.3.2 Computing Power

The brain and in general biology are extremely energy efficient. This contrasts with the large power budgets of today's computers. Many successful approaches in machine learning rely on very large models, requiring supercomputers or GPU clusters, which draw massive amounts of energy. Therefore, researchers are exploring biologically inspired methods of energy-efficient computation (Krichmar et al., 2019).

As discussed in chapter 5, principles of efficient coding in neuroscience can help us understand biology's natural efficiency, providing a framework for energy-efficient designs. For instance, neurons are asynchronous (i.e., not driven by a master clock) and event driven. Brain networks are massively parallel and stay in a low-power mode when not firing or receiving an action potential. Neuromorphic computing reflects this approach yielding hardware that utilize orders of magnitude less power than conventional computer architectures. One of the neurorobotic design principles introduced in chapter 5 was a multitasking event-driven architecture to allow robots to be more responsive to dynamic environments. It was shown that neurorobotic models of the basal ganglia following this architecture led to robots that rapidly selected actions corresponding to current stimulus events and the agent's needs. Such a just-in-time system may yield computing savings. Lastly, embodiment and morphological computation can save energy by allowing the body's layout of sensors and actuators, as well as its morphology, to handle processing, thus alleviating the power needs of the computer or brain.

10.3.3 Knowledge and Data Scarcity

Knowledge and data scarcity is a longstanding challenge in artificial intelligence. There are examples of humans being able to learn complex tasks with a very small amount of training input, in a phenomenon known as **poverty of the stimulus**. An example of this is the ability of children to learn languages and form sentences never heard before by hearing only a limited number of examples. On the other hand, machine learning algorithms typically require a vast number of training examples. For certain domains, such as image classification or natural language processing, there may be enough data to achieve success with current machine learning approaches. However, domains involving rare occurrences or sparse rewards face far more challenges in obtaining sufficient data. In addition, the challenges of energy consumption described previously intensify as data requirements increase.

Knowledge and data scarcity may be addressed by looking at the learning behaviors of biological organisms. Chapter 7 described several behavioral trade-offs that force the agent to work with a limited number of resources and make decisions on which types of data or knowledge are worth obtaining. Motivation and needs drive the agent to maximize a specific objective at any given time. Although not all artificial intelligence applications facing knowledge and data scarcity are in a physical environment, neurorobotic navigation strategies discussed in chapter 8 are applicable for efficient information foraging.

10.3.4 Human-Compatible Interactions

Many uses of artificial intelligence and robotics are isolated from people, perhaps occasionally gathering inputs from people and reporting outputs. Interaction with humans requires continuous communication and adaptation, not only producing the correct results but also ensuring that the interaction is intuitive and naturalistic. The **Turing test**, originally called the imitation game by Alan Turing in the 1950s, tested machine intelligence by having a human interact with the machine and judge how *human* the machine appears to be by asking questions. As depicted in figure 10.1, if the human is unable to tell whether the machine is human or not, the machine is considered to have passed the test. The test continues to be relevant today, particularly in natural language processing.

The principles of the Turing test also apply to robotics. However, the interaction has more emphasis on the robot's physical interactions rather than just its speech. In human-robot interactions it is crucial to earn and maintain the trust of the human. Beyond avoiding obviously harmful actions trust can be developed when the thought processing of the robot is explainable and beneficial to the human. Current approaches to human-robot interactions rely on design experts to ensure that the behavioral outputs of a robot are compatible with humans. However, this can lead to unnatural interactions and inflexibility.

The neurorobotic examples of social affect and language described in chapter 9 may serve as a good start toward developing human-compatible systems. Neurorobots can lead to a better understanding or use of emotions in social situations. Communication may be more natural because neurorobotics emphasizes grounding language to actions in the real world. By basing neurorobots on neurobiological studies of animal and human behavior the cognitive processes may be more explainable to humans. These aspects of neurorobotics could enable robots to communicate and interact more like humans.

10.4 Neurorobotics in Five, Ten, and Twenty Years

As we have seen, neurorobotics is not only an interesting academic field but also a technologically relevant one that can help us address longstanding challenges in artificial

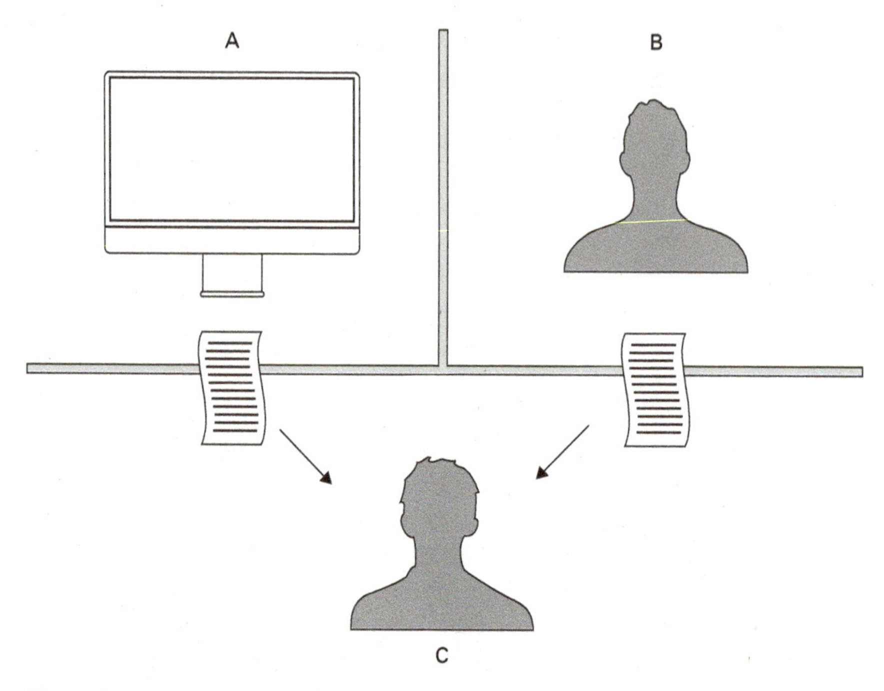

Figure 10.1
Diagram of the basic Turing test. A human (*C*) communicates with a computer (*A*) and another human (*B*) via text-based chat. None of the agents can see each other or communicate by any other means. The human evaluator must determine which of their conversational partners is the computer and which is the human. If the evaluator is unable to tell, the computer is said to have passed the Turing test. Adapted from https://en.wikipedia.org/wiki/Turing_test#/media/File:Turing_test_diagram.png.

intelligence and real-world applications. Despite the progress made so far, there is much more that neurorobotics can contribute toward future developments. Although it is risky to speculate where this field will be in the future, we briefly outline a prospective timeline for how neurorobotics can continue to grow and mature (see figure 10.2). Typically, unexpected events result in such predictions being wildly incorrect. However, we do believe that neurorobotics in some form will be a viable and impactful research area in the near and long terms.

10.4.1 Neurorobotics in Five Years

We expect that new neuroscience discoveries will inform neurorobots, and vice versa. Progress in learning and memory may lead to applications capable of continual learning. Advances in our understanding of multimodal sensory systems may be incorporated into neurorobotics that do not only classify but also understand the meaning of what they are perceiving. Given the recent achievements of artificial intelligence, hybrid systems combining machine learning and deep learning with neurorobotic design principles could lead to interesting applications in autonomous driving, assistive robots, and manufacturing.

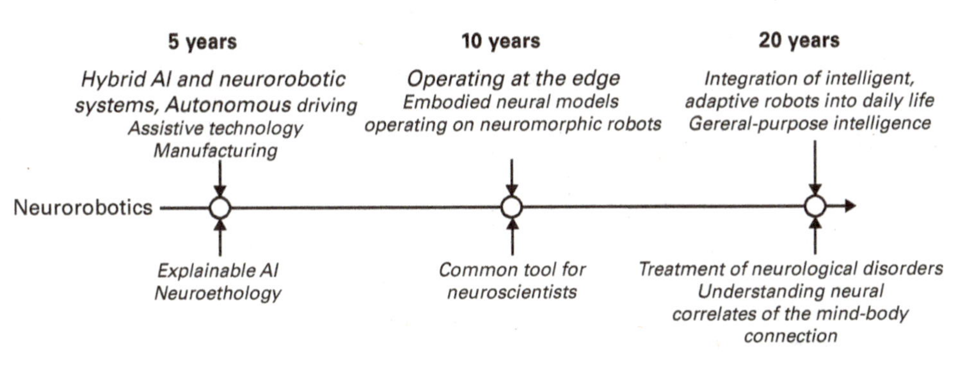

Figure 10.2
Projection of neurorobotics advances in the next five, ten, and twenty years. The *top row* describes potential neurorobotic advances in computing and technology. The *bottom row* describes potential neurorobotic advances in cognitive science and neuroscience.

The holistic neurorobotics approach may be able to advance our understanding of how neural activity leads to animal behavior. It has been suggested that current neuroscience research is removed from naturalistic behavior (Datta et al., 2019; Krakauer et al., 2017). They call for **neuroethology**, which is the science of quantifying naturalistic behaviors for understanding the brain. Although there has been incredible progress on the tools to examine and manipulate the nervous system, the controlled laboratory settings in which the brain is being observed is not the same as animal behavior in the wild. We have suggested that neurorobotics could be a form of neuroethology (Chen et al., 2020). A neuroroboticist can design experiments in which a robot is performing natural behavior, which has the advantage over animal models that the neuroroboticist can record the complete nervous system over the lifetime of the robot's behavior. Furthermore, mapping the neural network activity to observable behavior may be a means toward **explainable artificial intelligence**.

10.4.2 Neurorobotics in Ten Years

By ten years we believe that neurobiological concepts in learning and memory, navigation, decision making, and social behavior, to name a few, will have found their way into practical applications. Progress in neuromorphic computing and algorithms will lead to applications that can run at the **edge** with little human intervention. This may lead to advances in search and rescue robots and in robots capable of autonomously exploring unknown environments such as the deep sea or extraterrestrial planets.

At ten years neurorobotics could become an important component of neuroscience, where results can be directly compared with animal models under similar behavioral conditions. Although we suggested that neurorobotics could advance neuroscience within five years by being a form of ethology, we hope that neurorobotics, along with electrodes, brain imaging, and other tools, becomes a component of a neuroscientist's toolbox.

10.4.3 Neurorobotics in Twenty Years

As we approach twenty years we hope that neurorobotics will achieve more general intelligence rather than being designed for specific tasks. In fact, the delineation between conventional robotics and neurorobotics may be blurred, with all robots possessing some neurobiologically inspired aspects. With the rapid advances in computing and other technologies, it is hard to predict the inventions and breakthroughs twenty years forward. However, we do believe that neurorobotics and cognitive machines, in some form, will seamlessly be a part of our everyday lives.

Neurorobotics may be able to treat a range of disorders, including Alzheimer's, Parkinson's, and other neurological diseases. The new knowledge gained in cognitive science and neuroscience will allow the development of intelligent, adaptive robots in daily life that are genuinely beneficial to humanity. This might be in the form of intelligent brain-machine interfaces that adapt to the user's needs and augment brain responses. Neurorobots with human compatible interactions will be able to assist the elderly and disabled. These assistive robots will not be seen as something threatening but rather as a thoughtful companion. In addition, neurorobotics could shed light on some of the deeper cognitive science questions, such as the mind-body problem.

10.5 Summary and Conclusions

In this book we have studied how the brain and the body are intricately connected with each other and the environment. Neurorobotics allows us to study these interactions and test them in the real world on a physical system. Principles of neuroscience are used to improve robotics beyond traditional approaches, and likewise robots following the neurorobotic approach allows neuroscience theories to be tested and validated in an embodied environment. Developments in both fields continue to yield interesting results, inspiring further research directions on both sides and addressing some of the most difficult challenges and applications in society today.

It is our hope that this book will at the very least inspire readers to consider some of these concepts in their work and research. Optimistically, we hope that it inspires a new generation to continue advancing the exciting field of neurorobotics.

Glossary

Acetylcholine A neuromodulator important for increasing attention to important features, and decreasing attention to distractions.

Action potential How a neuron conveys a message or fires a spike. It involves a rapid increase in the neuron's voltage or membrane potential, followed by a rapid decrease in membrane potential to a resting level.

Actions The behavior that an agent can take for a given state.

Actuators Components that cause movement, such as motors or muscles.

Agent Broadly referring to any active entity, it can represent a human, animal, robot, or simulation.

Amygdala A brain area important for emotion and memory.

Artificial intelligence The simulation of intelligence through machines or software. It is a broad field that includes expert systems, neural networks, and machine learning.

Autonomous robots Machines that sense, think, and act without human intervention.

Axon Output of a neuron in the form of a cable that connects with other neurons.

Basal ganglia Brain area important for initiating or suppressing movements.

BCM learning rule Named after its originators: Bienenstock, Cooper, and Munro. A form of Hebbian learning with a sliding threshold that stabilizes the weights.

Catastrophic forgetting Forgetting previously learned information as a result of learning new information. This is an issue with many artificial learning systems.

Central nervous system (CNS) The brain and the spinal cord.

Cerebellum A brain structure important for fine tuning movements.

Classical conditioning An experimental psychology paradigm to make associations between salient stimuli and neutral stimuli.

Cognitive map A concept in which rats, humans, and other organisms have a mental map that can be used flexibly to navigate space or concepts.

Cognitive robotics A broad area of robotics that emphasizes biologically inspired behavior and intelligence through contributions from cognitive science, neuroscience, and psychology.

Complementary learning systems A memory theory in which the hippocampus rapidly learns new information that gradually becomes consolidated in the cortex.

Conditioned response The response driven by the conditioned stimulus after an association is made between the unconditioned stimulus and the conditioned stimulus.

Conditioned stimulus A neutral stimulus that is paired with an unconditioned stimulus.

Continual learning Learning multiple tasks in sequence without forgetting past tasks.

Contralateral connections Connections between neurons that cross from one side of the body to the other.

Cortex A brain structure with a repeating layered structure, important for processing sensory information and generating motor outputs.

Dead reckoning Estimating an agent's location by calculating its movements from a starting location.

Declarative memory Memory for places, events, and autobiographical information. Called declarative memory because these are memories that can be declared (verbally if by a human).

Dendrite Tree like structure of the neuron that receives synaptic input.

Dopamine A neuromodulator important for reward prediction.

Edge (edge computing) Systems and data processing that execute close to their use; for example, a mobile phone that is not using the cloud for information or a robot that is not connected to the internet.

Electroencephalogram (EEG) Measures brain-wave patterns with electrodes attached to the scalp.

Excitatory connection Connection in which the sending neuron increases the activity of the receiving neuron.

Expected uncertainty The known degree of unreliability of predictive relationships in the environment, proposed to drive activity within the cholinergic system.

Explainable artificial intelligence A research area that attempts to understand how artificial intelligence systems derive their decisions or outputs.

Exteroceptive sensors Sensors that obtain information from the external environment.

Extinction The decrease in learning or expected value when the salient stimulus is removed.

Figure-ground segregation The problem of separating an object (figure) from the background (ground). This challenge exists in natural and artificial visual systems.

Glucocorticoids Stress hormones important for triggering a *fight or flight* response.

Grid cell A neuron found in the entorhinal cortex that is active with a repeating, gridlike pattern across an environment.

Hebbian learning An unsupervised learning rule in which the weight between two co-active neurons is increased.

Hippocampal indexing theory Memories distributed in the neocortex are stored and recalled by indices in the hippocampus.

Hippocampus A brain area known to be important for learning and memory as well as for navigation.

Hormones Chemical messengers in the body that can affect the brain and other organs.

Hypothalamus A subcortical brain area that controls many bodily functions, such as body temperature, thirst, hunger, and sleep.

Inertial measurement unit (IMU) Calculates movement and current position with self-movement sensors such as an accelerometer, compass, and gyroscope.

Inhibitory connection Connection in which the sending neuron decreases the activity of the receiving neuron.

Interoceptive sensors Sensors that monitor the internal system.

Intrinsic motivation Obtaining value for its own sake, such as novelty seeking or play, rather than to satisfy some need.

Ipsilateral connections Connections between neurons that stay on the same side of the body.

Latent learning Acquiring knowledge without rewards and leveraging previous learned knowledge when the payoff comes later.

Lifelong learning Learning and reusing information over the course of a lifetime.

Long-term depression (LTD) A long-lasting decrease in the synaptic efficacy between two neurons.

Long-term potentiation (LTP) A long-lasting increase in the synaptic efficacy between two neurons.

Markov decision process A formalization to represent the best and most probable actions to take for a given environmental state.

Markov localization A probabilistic method to estimate one's location given the current sensory readings.

Mean firing-rate neuron A neuron model that simulates the average firing rate of a population of neurons.

Membrane potential The difference in electrical potential between the inside of the neuron and the outside of the neuron.

Mental simulation Thinking through possible actions and outcomes without taking those actions.

Mirror neurons Neurons in the premotor cortex that are active when one observes another's movements.

Model-based reinforcement learning Reward learning paradigm that creates a model of the environment, which can be in the form of transition probabilities between states. The model is then used to choose the best action for a given state.

Model-free reinforcement learning Reward learning paradigm that attempts to choose the best action for a given state.

Morphological computation Certain processes are performed by the body that otherwise would need to be performed by the brain. This reduces the computation load on the brain.

Morris water maze A test of spatial memory in which a rat must swim through opaque water until it finds a platform hidden beneath the surface.

Motor efference copy A copy of a motor command that is fed back to the nervous system. The motor efference copy can be used to verify that the movement had the desired outcome.

Multimodal Receiving or processing more than one sensory modality (e.g., auditory, and visual).

Neocortex The portion of the cortex that has six cellular layers.

Neuroethology Studying natural behaviors to understand the brain.

Neuromodulators Neurochemical signals that can have a broad, long-lasting effect on the synapse or neural activity.

Neuromorphic chip Computer hardware that mimics the neurons and synapses in the nervous system, typically designed as spiking neurons that can operate in parallel.

Neurons Cells or components that integrate information and output an activity level.

Neurorobots Robots whose control has been modeled after some aspect of the brain and nervous system.

Neurotransmitter A chemical released from the presynaptic neuron that can have an excitatory or inhibitory effect on the postsynaptic neuron.

Noradrenaline or **norepinephrine** A neuromodulator important for signaling unexpected events.

Obstacle avoidance A reactive path-planning response to detect and circumvent an obstacle that impedes the agent's progress.

Occupancy grid Divides a space into a grid or matrix that denotes how often the agent is in each grid location or whether the grid location is open and traversable.

Optic flow The pattern of movement across one's visual field, which can result from moving objects or from one's own movement.

Oxytocin A hormone important for forming social and maternal bonds.

Path integration Updating one's position based on the distance and direction of prior movements.

Perceptual categorization Learning categories or classes of information from one's experience.

Peripheral nervous system (PNS) Handles sensory information from the internal (i.e., within the body) and external (i.e., outside the body) environments.

Place cells A neuron found typically in the hippocampus that is active in a specific location of an environment.

Policy A description of the action plans that an agent has for a set of states and payoffs.

Postsynaptic neuron The receiving neuron side of a synapse.

Poverty of the stimulus The idea that humans are exposed to relatively little data during development in proportion to their rich linguistic and cognitive capabilities.

Presynaptic neuron The sending neuron side of a synapse.

Procedural memory Memory for learning a skill (e.g., a tennis swing). The memory is recalled as an action but may not necessarily be described or declared.

Receptive field The area of sensory space that a neuron samples and responds.

Recurrent neural network A neural network that has high connectivity between neurons within the same layer.

Reinforcement learning Learning occurs through positive or negative feedback from the learner's experience in the environment.

Rescorla-Wagner rule A mathematical formulation used to model classical conditioning; it can account for many experimental psychology phenomena.

Reward The payoff in reinforcement learning when the agent reaches a certain state. The reward can be a positive or negative value.

Reward prediction error The difference between the expected reward and the actual reward.

Schema A set of items or actions bound together by a common context.

Sensors Components that receive stimuli (such as light, sound, and vibration) from the environment.

Serotonin A neuromodulator important for harm aversion or impulsiveness.

Signal-to-noise ratio The response to the salient signal divided by the response to distractions.

Simultaneous localization and mapping algorithms (SLAM) A methodology from robot navigation that builds a map as the agent experiences the environment and localizes itself on this map.

Small-world network An architecture in which any two nodes in the network can be reached by a relatively small number of connections.

Softmax function An action selection method that turns a set of values into a probability distribution. The temperature parameter in the softmax function can change the shape of the distribution, with low values making the distribution flatter (explore) and high values making the distribution peakier (exploit).

Soma The cell body of a neuron that integrates inputs.

Spike-timing dependent plasticity (STDP) The timing of the presynaptic and postsynaptic action potentials is used to dictate the amount of weight change between two neurons.

Spiking neuron A neuron model that simulates the action potential with a spike.

State The agent's environmental or sensory condition at a given time.

Superior colliculus A midbrain area that controls rapid eye movements known as saccades.

Supervised learning Plasticity driven by a teacher or some external signal to help determine what is good or not good.

Synapses Connections between neurons.

Thalamus A brain area that processes and relays information between the cortex and the spinal cord as well as other subcortical brain areas.

Theory of mind An understanding of the beliefs, desires, and intents of another agent.

Turing test A test designed by Alan Turing for evaluating machine intelligence.

Unconditioned response The innate response driven by the unconditioned stimulus.

Unconditioned stimulus A stimulus that has preexisting valence or saliency.

Unexpected uncertainty Large changes in the environment that violate prior expectations. Proposed to drive activity within the noradrenergic system.

Unsupervised learning Plasticity driven by co-activity rather than by some extrinsic teacher or supervisory signal.

Value iteration Starting from the current state, iteration proceeds through the different states using transition probabilities to estimate the best course of action. Used in model-based reinforcement learning.

Value system A basic assumption of what is good, bad, or interesting for an agent. Value systems are used to shape behavior.

Weight normalization A strategy to keep synaptic weights within a range.

References

Abbott, J. T., J. L. Austerweil, and T. L. Griffiths. (2015). Random walks on semantic networks can resemble optimal foraging. *Psychological Review* 122:558–569.

Abbott, L. F., and P. Dayan. 2005. *Theoretical Neuroscience*. Cambridge, MA: MIT Press.

Abraham, W. C., and A. Robins. 2005. Memory retention—The synaptic stability versus plasticity dilemma. *Trends in Neurosciences* 28:73–78.

Alyan, S., and B. L. McNaughton. 1999. Hippocampectomized rats are capable of homing by path integration. *Behavioral Neuroscience* 113:19–31.

Aponte, Y., D. Atasoy, and S. M. Sternson. 2011. AgRP neurons are sufficient to orchestrate feeding behavior rapidly and without training. *Nature Neuroscience* 14:351–355.

Arkin, R. C. 1998. *Behavior-Based Robotics*. Cambridge, MA: MIT Press.

Arkin, R. C. 2016. Ethics and autonomous systems: Perils and promises [Point of view]. *Proceedings of the IEEE* 104:1779–1781.

Asada, M., K. Hosoda, Y. Kuniyoshi, H. Ishiguro, T. Inui, Y. Yoshikawa, M. Ogino, and C. Yoshida. 2009. Cognitive developmental robotics: A survey. *IEEE Transactions on Autonomous Mental Development* 1:12–34.

Asher, D. E., A. Zaldivar, B. Barton, A. A. Brewer, and J. L. Krichmar. 2012. Reciprocity and retaliation in social games with adaptive agents. *IEEE Transactions on Autonomous Mental Development* 4:226–238.

Asher, D. E., A. Zaldivar, and J. L. Krichmar. 2010. Effect of neuromodulation on performance in game playing: A modeling study. In *2010 IEEE 9th International Conference on Development and Learning*. New York: IEEE.

Asimov, I. 1950. *I, Robot*. New York: Doubleday.

Aston-Jones, G., and J. D. Cohen. 2005. An integrative theory of locus coeruleus-norepinephrine function: Adaptive gain and optimal performance. *Annual Review of Neuroscience* 28:403–450.

Avery, M. C., and J. L. Krichmar. 2017. Neuromodulatory systems and their interactions: A review of models, theories, and experiments. *Frontiers in Neural Circuits* 11, Article 108.

Avery, M. C., D. A. Nitz, A. A. Chiba, and J. L. Krichmar. 2012. Simulation of cholinergic and noradrenergic modulation of behavior in uncertain environments. *Frontiers in Computational Neuroscience* 6, Article 5.

Balkenius, C., T. A Tjøstheim, and B. Johansson. 2018. Arousal and awareness in a humanoid robot. In *CEUR Workshop Proceedings 2287 (2018)*. Lund, Sweden: Lund University Publications. http://ceur-ws.org/Vol-2287/paper25.pdf.

Ball, D., S. Heath, J. Wiles, G. Wyeth, P. Corke, and M. Milford 2013. OpenRatSLAM: An open source brain-based SLAM system. *Autonomous Robots* 34:149–176.

Barraquand, J., B. Langlois, and J. Latombe. 1991. Numerical potential field techniques for robot path planning. In *Proceedings of the Fifth International Conference on Advanced Robotics 'Robots in Unstructured Environments*. New York: IEEE.

Bassett, D. S., and E. T. Bullmore. 2017. Small-world brain networks revisited. *Neuroscientist* 23:499–516.

Bechara, A., H. Damasio, A. R. Damasio, and G. P. Lee. 1999. Different contributions of the human amygdala and ventromedial prefrontal cortex to decision-making. *Journal of Neuroscience* 19:5473–5481.

Bekey, G. A. 2017. *Autonomous Robots: From Biological Inspiration to Implementation and Control*. Cambridge, MA: MIT Press.

Belkaid, M., and J. L. Krichmar. 2020. Modeling uncertainty-seeking behavior mediated by cholinergic influence on dopamine. *Neural Networks* 125:10–18.

Ben-Ami Bartal, I., J. Decety, and P. Mason. 2011. Empathy and pro-social behavior in rats. *Science* 334: 1427–1430.

Beyeler, M., N. Oros, N. Dutt, and J. L. Krichmar. 2015. A GPU-accelerated cortical neural network model for visually guided robot navigation. *Neural Networks* 72:75–87.

Beyeler, M., M. Richert, N. D. Dutt, and J. L. Krichmar. 2014. Efficient spiking neural network model of pattern motion selectivity in visual cortex. *Neuroinformatics* 12:435–454.

Beyeler, M., E. L. Rounds, K. D. Carlson, N. Dutt, and J. L. Krichmar. 2019. Neural correlates of sparse coding and dimensionality reduction. *PLoS Computational Biology* 15:e1006908.

Bhounsule, P. A., J. Cortell, A. Grewal, B. Hendriksen, J.G.D. Karssen, C. Paul, and A. Ruina. 2014. Low-bandwidth reflex-based control for lower power walking: 65 km on a single battery charge. *International Journal of Robotics Research* 33:1305–1321.

Bi, G. Q., and M. M. Poo. 1998. Synaptic modifications in cultured hippocampal neurons: Dependence on spike timing, synaptic strength, and postsynaptic cell type. *Journal of Neuroscience* 18:10464–10472.

Bienenstock, E. L., L. N. Cooper, and P. W. Munro. 1982. Theory for the development of neuron selectivity: Orientation specificity and binocular interaction in visual cortex. *Journal of Neuroscience* 2:32–48.

Billard, A., and M. J. Mataric. 2001. Learning human arm movements by imitation: Evaluation of a biologically inspired connectionist architecture. *Robotics and Autonomous Systems* 37:145–160.

Bologna, L. L., J. Pinoteau, R. Brasselet, M. Maggiali, and A. Arleo. 2011. Encoding/decoding of first and second order tactile afferents in a neurorobotic application. *Journal of Physiology-Paris* 105:25–35.

Bongard, J., V. Zykov, and H. Lipson. 2006. Resilient machines through continuous self-modeling. *Science* 314:1118–1121.

Borenstein, J., and R. Arkin. 2016. Robotic nudges: The ethics of engineering a more socially just human being. *Science and Engineering Ethics* 22:31–46.

Bossi, F., C. Willemse, J. Cavazza, S. Marchesi, V. Murino, and A. Wykowska. 2020. The human brain reveals resting state activity patterns that are predictive of biases in attitudes toward robots. *Science Robotics* 5. https://doi.org/10.1126/scirobotics.abb6652.

Boucenna, S., P. Gaussier, P. Andry, and L. Hafemeister. 2014. A robot learns the facial expressions recognition and face/non-face discrimination through an imitation game. *International Journal of Social Robotics* 6:633–652.

Boucenna, S., P. Gaussier, and L. Hafemeister. 2014. Development of first social referencing skills: Emotional interaction as a way to regulate robot behavior. *IEEE Transactions on Autonomous Mental Development* 6:42–55.

Boureau, Y. L., and P. Dayan. 2011. Opponency revisited: Competition and cooperation between dopamine and serotonin. *Neuropsychopharmacology: Official Publication of the American College of Neuropsychopharmacology* 36:74–97.

Bouret, S., and S. J. Sara. 2005. Network reset: A simplified overarching theory of locus coeruleus noradrenaline function. *Trends in Neurosciences* 28:574–582.

Braitenberg, V. 1986. *Vehicles: Experiments in Synthetic Psychology*. Cambridge, MA: MIT Press.

Brette, R., and W. Gerstner. 2005. Adaptive exponential integrate-and-fire model as an effective description of neuronal activity. *Journal of Neurophysiology* 94:3637–3642.

Brooks, R. A. 1986. A robust layered control-system for a mobile robot. *IEEE Journal on Robotics and Automation* 2:14–23.

Brooks, R. A. 1991. Intelligence without representation. *Artificial Intelligence* 47:139–159.

Cañamero, D. 1997. Modeling motivations and emotions as a basis for intelligent behavior. In *Proceedings of the First International Joint Conference on Autonomous Agents*. New York: Association for Computing Machinery, 148–155.

Canamero, L., A. J. Blanchard, and J. Nadel. 2006. Attachment bonds for human-like robots. *International Journal of Humanoid Robotics* 3:301–320.

Cangelosi, A., and M. Asada, eds. 2022. *Cognitive Robotics.* Cambridge, MA: MIT Press.

Cangelosi, A., and M., Schlesinger. 2015. *Developmental Robotics: From Babies to Robots* Cambridge, MA: MIT Press.

Cardinal, R. N., J. A. Parkinson, J. Hall, and B. J. Everitt. 2002. Emotion and motivation: The role of the amygdala, ventral striatum, and prefrontal cortex. *Neuroscience and Biobehavioral Reviews* 26:321–352.

Carey, R.J., E. N. Damianopoulos, and A. B. Shanahan. 2008. Cocaine effects on behavioral responding to a novel object placed in a familiar environment. *Pharmacology, Biochemistry and Behavior* 88:265–271.

Carr, L., M. Iacoboni, M. C. Dubeau, J. C. Mazziotta, and G. L. Lenzi. 2003. Neural mechanisms of empathy in humans: A relay from neural systems for imitation to limbic areas. *Proceedings of the National Academy of Sciences of the United States of America* 100:5497–5502.

Chen, K., T. Hwu, H. J. Kashyap, J. L. Krichmar, K. Stewart, J. Xing, and X. Zou. 2020. Neurorobots as a means toward neuroethology and explainable AI. *Frontiers in Neurorobotics* 14. https://doi.org/10.3389/fnbot.2020.570308.

Chersi, F. 2012. Learning through imitation: A biological approach to robotics. *IEEE Transactions on Autonomous Mental Development* 4:204–214.

Chiba, A. A., and J. L. Krichmar. 2020. Neurobiologically inspired self-monitoring systems. *Proceedings of the IEEE* 108:976–986.

Chiel, H. J., and R. D. Beer. 1997. The brain has a body: Adaptive behavior emerges from interactions of nervous system, body and environment. *Trends in Neurosciences* 20:553–557.

Choset, H., K. M. Lynch, S. Hutchinson, G. A. Kantor, W. Burgard, L. E. Kavraki, and S. Thrun. 2005. *Principles of Robot Motion: Theory, Algorithms, and Implementations.* Cambridge, MA: MIT Press.

Chou, T. S., L. D. Bucci, and J. L. Krichmar. 2015. Learning touch preferences with a tactile robot using dopamine modulated STDP in a model of insular cortex. *Frontiers in Neurorobotics* 9, Article 6.

Clark, A. 2013. Whatever next? Predictive brains, situated agents, and the future of cognitive science. *Behavioral and Brain Sciences* 36:181–204.

Collins, S., A. Ruina, R. Tedrake, and M. Wisse. 2005. Efficient bipedal robots based on passive-dynamic walkers. *Science* 307:1082–1085.

Collins, S. H., and A. Ruina, A. 2005. A bipedal walking robot with efficient and human-like gait. In *Proceedings of the 2005 IEEE International Conference on Robotics and Automation.* New York: IEEE.

Cooper, L. N., and M. F. Bear. 2012. The BCM theory of synapse modification at 30: Interaction of theory with experiment. *Nature Reviews Neuroscience* 13:798–810.

Cox, B. R., and J. L. Krichmar. 2009. Neuromodulation as a robot controller: A brain inspired design strategy for controlling autonomous robots. *IEEE Robotics & Automation Magazine* 16:72–80.

Cullen, K. E., and J. S. Taube. 2017. Our sense of direction: Progress, controversies, and challenges. *Nature Neuroscience* 20:1465–1473.

Cully, A., J. Clune, D. Tarapore, and J.-B. Mouret. 2015. Robots that can adapt like animals. *Nature* 521: 503–507.

Cuperlier, N., M. Quoy, and P. Gaussier. 2007. Neurobiologically inspired mobile robot navigation and planning. *Frontiers in Neurorobotics* 1, Article 3.

Datta, S. R., D. J. Anderson, K. Branson, P. Perona, and A. Leifer. 2019. Computational neuroethology: A call to action. *Neuron* 104:11–24.

Davies, M., N. Srinivasa, T.-H. Lin, G. Chinya, Y. Cao, S. H. Choday, G. Dimou, P. Joshi, N. Imam, S. Jain, et al. 2018. Loihi: A neuromorphic manycore processor with on-chip learning. *IEEE Micro* 38:82–99.

Daw, N. D., S. J. Gershman, B. Seymour, P. Dayan, and R. J. Dolan. 2011. Model-based influences on humans' choices and striatal prediction errors. *Neuron* 69:1204–1215.

Daw, N. D., S. Kakade, and P. Dayan. 2002. Opponent interactions between serotonin and dopamine. *Neural Networks* 15:603–616.

Derdikman, D., and E. I. Moser. 2010. A manifold of spatial maps in the brain. *Trends in Cognitive Sciences* 14:561–569.

Dolen, G., A. Darvishzadeh, K. W. Huang, and R. C. Malenka. 2013. Social reward requires coordinated activity of nucleus accumbens oxytocin and serotonin. *Nature* 501:179–184.

Dominey, P. F. 2013. Recurrent temporal networks and language acquisition—From corticostriatal neurophysiology to reservoir computing. *Frontiers in Psychology* 4, Article 500.

Dragoi, G., and S. Tonegawa. 2011. Preplay of future place cell sequences by hippocampal cellular assemblies. *Nature* 469:397–401.

Edelman, G. M. 1993. Neural Darwinism: Selection and reentrant signaling in higher brain function. *Neuron* 10:115–125.

Edelman, G. M., and J. A. Gally. 2001. Degeneracy and complexity in biological systems. *Proceedings of the National Academy of Sciences of the United States of America* 98:13763–13768.

Eichenbaum, H., and N. J. Cohen. 2014. Can we reconcile the declarative memory and spatial navigation views on hippocampal function? *Neuron* 83:764–770.

Fajen, B. R., and W. H. Warren. 2003. Behavioral dynamics of steering, obstacle avoidance, and route selection. *Journal of Experimental Psychology: Human Perception and Performance* 29:343–362.

Fields, R. D. 2015. A new mechanism of nervous system plasticity: Activity-dependent myelination. *Nature Reviews Neuroscience* 16:756–767.

Fiore, V. G., V. Sperati, F. Mannella, M. Mirolli, K. Gurney, K. Friston, R. J. Dolan, and G. Baldassarre. 2014. Keep focusing: Striatal dopamine multiple functions resolved in a single mechanism tested in a simulated humanoid robot. *Frontiers in Psychology* 5, Article 124.

Fischl, K. D., K. Fair, W. Tsai, J. Sampson, and A. Andreou. 2017. Path planning on the TrueNorth neurosynaptic system. In *Proceedings of the 2017 IEEE International Symposium on Circuits and Systems (ISCAS)*. New York: IEEE.

Fitzpatrick, P., and G. Metta. 2003. Grounding vision through experimental manipulation. *Philosophical Transactions of the Royal Society of London, Series A: Mathematical, Physical and Engineering Sciences* 361:2165–2185.

Fleischer, J. G., J. A. Gally, G. M. Edelman, and J. L. Krichmar. 2007. Retrospective and prospective responses arising in a modeled hippocampus during maze navigation by a brain-based device. *Proceedings of the National Academy of Sciences of the United States of America* 104:3556–3561.

Fleischer, J. G., B. Szatmary, D. Hutson, D. A. Moore, J. A. Snook, G. M. Edelman, and J. L. Krichmar. 2006. A neurally controlled robot competes and cooperates with humans in Segway soccer. In *Proceedings of the IEEE International Conference on Robotics and Automation*. New York: IEEE.

Fonio, E., Y. Benjamini, and I. Golani. 2009. Freedom of movement and the stability of its unfolding in free exploration of mice. *Proceedings of the National Academy of Sciences of the United States of America* 106: 21335–21340.

Foster, D. J., R. G. Morris, and P. Dayan. 2000. A model of hippocampally dependent navigation, using the temporal difference learning rule. *Hippocampus* 10:1–16.

Frank, M. J., and E. D. Claus. 2006. Anatomy of a decision: Striato-orbitofrontal interactions in reinforcement learning, decision making, and reversal. *Psychological Review* 113:300–326.

Friston, K. 2010. The free-energy principle: A unified brain theory? Nature Reviews *Neuroscience* 11:127–138.

Friston, K. J., G. Tononi, G. N. Reeke Jr., O. Sporns, and G. M. Edelman. 1994. Value-dependent selection in the brain: Simulation in a synthetic neural model. *Neuroscience* 59:229–243.

Fuster, J. M. 2004. Upper processing stages of the perception-action cycle. *Trends in Cognitive Sciences* 8:143–145.

Galindo, C., A. Saffiotti, S. Coradeschi, P. Buschka, J. A. Fernandez-Madrigal, and J. Gonzalez. 2005. Multi-hierarchical semantic maps for mobile robotics. In *Proceedings of the 2005 IEEE/RSJ International Conference on Intelligent Robots and Systems*. New York: IEEE.

Galluppi, F., C. Denk, M. C. Meiner, T. C. Stewart, L. A. Plana, C. Eliasmith, S. Furber, and J. Conradt. 2014. Event-based neural computing on an autonomous mobile platform. In *Proceedings of the 2014 IEEE International Conference on Robotics and Automation (ICRA)*. New York: IEEE.

George, D., and J. Hawkins. 2009. Towards a mathematical theory of cortical micro-circuits. *PLoS Computational Biology* 5:e1000532.

Glascher, J., N. Daw, P. Dayan, and J. P. O'Doherty. 2010. States versus rewards: Dissociable neural prediction error signals underlying model-based and model-free reinforcement learning. *Neuron* 66:585–595.

Grossberg, S. 2013. Adaptive resonance theory: How a brain learns to consciously attend, learn, and recognize a changing world. *Neural Networks* 37:1–47.

Grossberg, S. 2017. Towards solving the hard problem of consciousness: The varieties of brain resonances and the conscious experiences that they support. *Neural Networks* 87:38–95.

Gurney, K., T. J. Prescott, and P. Redgrave. 2001a. A computational model of action selection in the basal ganglia. I. A new functional anatomy. *Biological Cybernetics* 84:401–410.

Gurney, K., T. J. Prescott, and P. Redgrave. 2001b. A computational model of action selection in the basal ganglia. II. Analysis and simulation of behaviour. *Biological Cybernetics* 84:411–423.

Hahn, T. M., J. F. Breininger, D. G. Baskin, and M. W. Schwartz. 1998. Coexpression of AgRP and NPY in fasting-activated hypothalamic neurons. *Nature Neuroscience* 1:271–272.

Hasselmo, M. E., and J. McGaughy. 2004. High acetylcholine levels set circuit dynamics for attention and encoding and low acetylcholine levels set dynamics for consolidation. *Progress in Brain Research* 145:207–231.

Hawkins, J. 2017. What intelligent machines need to learn from the neocortex. *IEEE Spectrum Magazine*, 35–40.

Hebb, D. O. 1949. *The Organization of Behavior: A Neuropsychological Theory*. New York: Wiley.

Heisler, L. K., H. M. Chu, T. J. Brennan, J. A. Danao, P. Bajwa, L. H. Parsons, and L. H. Tecott. 1998. Elevated anxiety and antidepressant-like responses in serotonin 5-HT1A receptor mutant mice. *Proceedings of the National Academy of Sciences of the United States of America* 95:15049–15054.

Hesslow, G. 2012. The current status of the simulation theory of cognition. *Brain Research* 1428:71–79.

Hickok, G. 2009. Eight problems for the mirror neuron theory of action understanding in monkeys and humans. *Journal of Cognitive Neuroscience* 21:1229–1243.

Hickok, G., and M. Hauser. 2010. (Mis)understanding mirror neurons. *Current Biology* 20:R593–594.

Hickok, G., J. Houde, and F. Rong. 2011. Sensorimotor integration in speech processing: Computational basis and neural organization. *Neuron* 69:407–422.

Hiolle, A., K. A. Bard, and L. Canamero. 2009. Assessing human reactions to different robot attachment profiles. In *Proceedings of Ro-Man 2009: The 18th IEEE International Symposium on Robot and Human Interactive Communication*, V. 1 and 2, 824–829. New York: Association for Computing Machinery.

Hiolle, A., L. Canamero, M. Davila-Ross, and K. A. Bard. 2012. Eliciting caregiving behavior in dyadic human-robot attachment-like interactions. *ACM Transactions on Interactive Intelligent Systems* 2:1–24.

Hok, V., E. Save, P. P. Lenck-Santini, and B. Poucet. 2005. Coding for spatial goals in the prelimbic/infralimbic area of the rat frontal cortex. *Proceedings of the National Academy of Sciences of the United States of America* 102:4602–4607.

Holland, O. 2003. Exploration and high adventure: The legacy of Grey Walter. *Philosophical Transactions of the Royal Society of London, Series A: Mathematical, Physical and Engineering Sciences* 361:2085–2121.

Hwang, J., J. Kim, A. Ahmadi, M. Choi, and J. Tani. 2020. Dealing with large-scale spatio-temporal patterns in imitative interaction between a robot and a human by using the predictive coding framework. *IEEE Transactions on Systems, Man, and Cybernetics: Systems* 50:1918–1931.

Hwu, T., J. Isbell, N. Oros, and J. Krichmar. 2017. A self-driving robot using deep convolutional neural networks on neuromorphic hardware. In *Proceedings of the 2017 IEEE International Joint Conference on Neural Networks (IJCNN)*, 635–641. New York: IEEE.

Hwu, T., H. J. Kashyap, and J. L. Krichmar. 2020. A neurobiological schema model for contextual awareness in robotics. In *Proceedings of the 2020 IEEE International Joint Conference on Neural Networks (IJCNN)*. New York: IEEE.

Hwu, T., J. Krichmar, and X. Zou. 2017. A complete neuromorphic solution to outdoor navigation and path planning. In *Proceedings of the 2017 IEEE International Symposium on Circuits and Systems (ISCAS)*, 2707–2710. New York: IEEE.

Hwu, T., and J. L. Krichmar. 2020. A neural model of schemas and memory encoding. *Biological Cybernetics* 114:169–186.

Hwu, T., A. Y. Wang, N. Oros, and J. L. Krichmar. 2018. Adaptive robot path planning using a spiking neuron algorithm with axonal delays. *IEEE Transactions on Cognitive and Developmental Systems* 10:126–137.

Iida, F., and R. Pfeifer. 2004. Self-stabilization and behavioral diversity of embodied adaptive locomotion. In *Embodied Artificial Intelligence*, edited by F. Iida, R. Pfeifer, L. Steels, and Y, Kuniyoshi. Lecture Notes in Computer Science, V. 3139. Berlin: Springer.

Izhikevich, E. M. (2004). Which model to use for cortical spiking neurons? *IEEE Transactions on Neural Networks* 15:1063–1070.

Jasinska, A. J., C. A. Lowry, and M. Burmeister. 2012. Serotonin transporter gene, stress and raphe-raphe interactions: A molecular mechanism of depression. *Trends in Neurosciences* 35:395–402.

Johansson, B., and C. Balkenius. 2018. A computational model of pupil dilation. *Connection Science* 30:5–19.

Johansson, B., T. A. Tjøstheim, and C. Balkenius. 2020. Epi: An open humanoid platform for developmental robotics. *International Journal of Advanced Robotic Systems* 17:1729881420911498.

Kirby, R., J. Forlizzi, and R. Simmons. 2010. Affective social robots. *Robotics and Autonomous Systems* 58:322–332.

Kostavelis, I., and A. Gasteratos. 2017. Semantic maps from multiple visual cues. *Expert Systems with Applications* 68:45–57.

Kotseruba, I., and Tsotsos, J. K. (2018). A review of 40 years in cognitive architecture research core cognitive abilities and practical applications. https://arxiv.org/abs/1610.08602v.

Kountouriotis, G. K., K. A. Shire, C. D. Mole, P. H. Gardner, N. Merat, and R. M. Wilkie. 2013. Optic flow asymmetries bias high-speed steering along roads. *Journal of Vision* 13:23.

Krakauer, J. W., A. A. Ghazanfar, A. Gomez-Marin, M. A. MacIver, and D. Poeppel. 2017. Neuroscience needs behavior: Correcting a reductionist bias. *Neuron* 93:480–490.

Kravitz, D. J., K. S. Saleem, C. I. Baker, L. G. Ungerleider, and M. Mishkin. 2013. The ventral visual pathway: An expanded neural framework for the processing of object quality. *Trends in Cognitive Sciences* 17:26–49.

Krichmar, J. L. 2008. The neuromodulatory system—A framework for survival and adaptive behavior in a challenging world. *Adaptive Behavior* 16:385–399.

Krichmar, J. L. 2013. A neurorobotic platform to test the influence of neuromodulatory signaling on anxious and curious behavior. *Frontiers in Neurorobotics* 7:1–17.

Krichmar, J. L. 2016. Path planning using a spiking neuron algorithm with axonal delays. In *Proceedings of the 2016 IEEE Congress on Evolutionary Computation*, 1219–1226. New York: IEEE.

Krichmar, J. L. 2018. Neurorobotics—A thriving community and a promising pathway toward intelligent cognitive robots. *Frontiers in Neurorobotics* 12:1–11.

Krichmar, J. L., and T. S. Chou. 2018. A tactile robot for developmental disorder therapy. In *Proceedings of the Technology, Mind, and Society Conference (Techmindsociety'18)*. New York: Association for Computing Machinery.

Krichmar, J. L., and G. M. Edelman. 2002. Machine psychology: Autonomous behavior, perceptual categorization, and conditioning in a brain-based device. *Cerebral Cortex* 12:818–830.

Krichmar, J. L., D. A. Nitz, J. A. Gally, and G. M. Edelman. 2005. Characterizing functional hippocampal pathways in a brain-based device as it solves a spatial memory task. *Proceedings of the National Academy of Sciences of the United States of America* 102:2111–2116.

Krichmar, J. L., A., K. Seth, D. A. Nitz, J. G. Fleischer, and G. M. Edelman. 2005. Spatial navigation and causal analysis in a brain-based device modeling cortical-hippocampal interactions. *Neuroinformatics* 3:197–221.

Krichmar, J. L., W. Severa, M. S. Khan, and J. L. Olds. 2019. Making BREAD: Biomimetic strategies for artificial intelligence now and in the future. *Frontiers in Neuroscience* 13, Article 666.

Kropff, E., J. E. Carmichael, M. B. Moser, and E. I. Moser. 2015. Speed cells in the medial entorhinal cortex. *Nature* 523:419–424.

Kumaran, D., D. Hassabis, and J. L. McClelland. 2016. What learning systems do intelligent agents need? Complementary learning systems theory updated. *Trends in Cognitive Sciences* 20:512–534.

Laughlin, S. B., and T, J. Sejnowski. 2003. Communication in neuronal networks. *Science* 301:870–1874.

LeDoux, J. E. 2000. Emotion circuits in the brain. *Annual Review of Neuroscience* 23:155–184.

Lones, J., and L. Canamero. 2013. Epigenetic adaptation through hormone modulation in autonomous robots. In *Proceedings of the 2013 IEEE Third Joint International Conference on Development and Learning and Epigenetic Robotics (ICDL)*. New York: IEEE.

Lones, J., M. Lewis, and L. Cañamero. 2018. A hormone-driven epigenetic mechanism for adaptation in autonomous robots. *IEEE Transactions on Cognitive and Developmental Systems* 10:445–454.

Luquet, S., F A. Perez, T. S. Hnasko, and R. D. Palmiter. 2005. NPY/AgRP neurons are essential for feeding in adult mice but can be ablated in neonates. *Science* 310683–310685.

Markoff, J. 2015. Relax, the Terminator is far away. *New York Times*, May 25, 2015.

Markram, H., J. Lubke, M. Frotscher, and B. Sakmann. 1997. Regulation of synaptic efficacy by coincidence of postsynaptic APs and EPSPs. *Science* 275:213–215.

Marocco, D., A. Cangelosi, K. Fischer, and T. Belpaeme. 2010. Grounding action words in the sensorimotor interaction with the world: Experiments with a simulated iCub humanoid robot. *Frontiers in Neurorobotics* 4, Article 7.

Mataric, M. J., and B. Scassellati. 2016. Socially assistive robotics. In *Springer Handbook of Robotics, 1973–1993*, edited by Bruno Siciliano. Cham, Switzerland: Springer International.

McClelland, J. L., B. L. McNaughton, and R. C. O'Reilly. 1995. Why there are complementary learning systems in the hippocampus and neocortex: Insights from the successes and failures of connectionist models of learning and memory. *Psychological Review* 102:19–457.

Mermillod, M., A. Bugaiska, and P. Bonin. 2013. The stability-plasticity dilemma: Investigating the continuum from catastrophic forgetting to age-limited learning effects. *Frontiers in Psychology* 4, Article 504.

Merolla, P. A., J. V. Arthur, R. Alvarez-Icaza, A. S. Cassidy, J. Sawada, F. Akopyan, B. L., Jackson, N. Imam, C. Guo, Y. Nakamura, et al. 2014. Artificial brains. A million spiking-neuron integrated circuit with a scalable communication network and interface. *Science* 345:668–673.

Merrick, K. 2017. Value systems for developmental cognitive robotics: A survey. *Cognitive Systems Research* 41:38–55.

Merrick, K. E. 2010a. A comparative study of value systems for self-motivated exploration and learning by robots. *IEEE Transactions on Autonomous Mental Development* 2:119–131.

Merrick, K. E. 2010b. Modeling behavior cycles as a value system for developmental robots. *Adaptive Behavior* 18:237–257.

Milford, M., A. Jacobson, Z. Chen, and G. Wyeth. 2016. RatSLAM: Using models of rodent hippocampus for robot navigation and beyond. In *Robotics Research: The 16th International Symposium ISSR*, edited by Masayuki Inaba and Peter Corke. Springer Tracts in Advanced Robotics, V. 114, 467–485. Berlin: Springer.

Milford, M. J., and G. F. Wyeth. 2008. Mapping a suburb with a single camera using a biologically inspired SLAM system. *IEEE Transactions on Robotics* 24:1038–1053.

Miller, P. 2018. *An Introductory Course in Computational Neuroscience*. Cambridge, MA: MIT Press.

Moualla, A., S. Boucenna, A. Karaouzene, D. Vidal, and P. Gaussier. 2018. Is it useful for a robot to visit a museum? *Paladyn, Journal of Behavioral Robotics* 9:374–390.

Movellan, J. R. 1991. Contrastive Hebbian learning in the continuous Hopfield model. In *Connectionist Models*, edited by D. S. Touretzky, J. L. Elman, T. J. Sejnowski, and G. E. Hinton, 10–17. San Mateo, CA: Morgan Kaufmann.

Murata, S., H. Arie, T. Ogata, S. Sugano, and J. Tani. 2014. Learning to generate proactive and reactive behavior using a dynamic neural network model with time-varying variance prediction mechanism. *Advanced Robotics* 28:1189–1203.

Naude, J., S. Tolu, M. Dongelmans, N. Torquet, S. Valverde, G. Rodriguez, S. Pons, U. Maskos, A. Mourot, F. Marti, et al. 2016. Nicotinic receptors in the ventral tegmental area promote uncertainty-seeking. *Nature Neuroscience* 19:471–478.

O'Keefe, J., and Nadel, L. 1978. *The Hippocampus as a Cognitive Map.* Oxford, UK: Oxford University Press.

Olson, J. M., K. Tongprasearth, and D. A. Nitz. 2017. Subiculum neurons map the current axis of travel. *Nature Neuroscience* 20:170–172.

Oros, N., and J. L. Krichmar. 2013. *Smartphone Based Robotics: Powerful, Flexible and Inexpensive Robots for Hobbyists, Educators, Students and Researchers.* CECS Technical Report 13–16. Irvine: University of California.

Oudeyer, P. 2006. *Self-Organization in the Evolution of Speech.* Oxford, UK: Oxford University Press.

Oudeyer, P., F. Kaplan, and V. V. Hafner. 2007. Intrinsic motivation systems for autonomous mental development. *IEEE Transactions on Evolutionary Computation* 11:265–286.

Oudeyer, P. Y., and F. Kaplan. 2007. What is intrinsic motivation? A typology of computational approaches. *Frontiers in Neurorobotics* 1, Article 6.

Padilla, S. L., J. Qiu, M. E. Soden, E. Sanz, C. C. Nestor, F. D. Barker, A. Quintana, L. Zweifel, O. K. Ronnekleiv, M. J. Kelly, et al. 2016. Agouti-related peptide neural circuits mediate adaptive behaviors in the starved state. *Nature Neuroscience* 19:734–741.

Pavlov, I. P. 1929. Lectures on conditioned reflexes: Twenty-five years of objective study of the higher nervous activity (behaviour) of animals. *Nature* 124:400–401.

Pfeifer, R., and J. Bongard. 2006. *How the Body Shapes the Way We Think: A New View of Intelligence.* Cambridge, MA: MIT Press.

Pfeiffer, B. E., and D. J. Foster. 2013. Hippocampal place-cell sequences depict future paths to remembered goals. *Nature* 497:74–79.

Piaget, J. 1971. The theory of stages in cognitive development. In *Measurement and Piaget,* edited by Donald Ross Green, pp. ix, 283-ix, 283. New York: McGraw-Hill.

Pollack, J. B. 1989. No harm intended: Marvin L. Minsky and Seymour A. Papert. *Perceptrons: An Introduction to Computational Geometry,* Expanded Edition. Cambridge, MA: MIT Press, 1988. Pp. 292. $12.50 (paper). *Journal of Mathematical Psychology* 33:358–365.

Prescott, T. J., F. M. Montes González, K. Gurney, M. D. Humphries, and P. Redgrave. 2006. A robot model of the basal ganglia: Behavior and intrinsic processing. *Neural Networks* 19:31–61.

Prescott, T. J., P. Redgrave, and K. Gurney. 1999. Layered control architectures in robots and vertebrates. *Adaptive Behavior* 7:99–127.

Purves, D., G. J. Augustine, D. Fitzpatrick, W. C. Hall, A.-S. LaMantia, R. D. Mooney, M. L. Platt, and L. E. White. 2017. *Neuroscience,* 6th edn. New York: Oxford University Press.

Quinn, L. K., L. P. Schuster, M. Aguilar-Rivera, J. Arnold, D. Ball, E. Gygi, S. Heath, J. Holt, D. J. Lee, J. Taufatofua, et al. 2018. When rats rescue robots. *Animal Behavior and Cognition* 5:368–379.

Raheem, F. A., and M. M. Badr. 2017. Development of modified path planning algorithm using artificial potential field (APF) based on PSO for factors optimization. *American Scientific Research Journal for Engineering, Technology, and Sciences* 37:316–328.

Renaudo, E., B. Girard, R. Chatila, and M. Khamassi. 2015. Respective advantages and disadvantages of model-based and model-free reinforcement learning in a robotics neuro-inspired cognitive architecture. *Procedia Computer Science* 71:178–184.

Rescorla, R. A., and A. R. Wagner. 1972. A theory of Pavlovian conditioning: Variations in the effectiveness of reinforcement and nonreinforcement. In *Classical Conditioning II,* edited by A. H. Black and W. F. Prokasy, 64–99. New York: Appleton-Century-Crofts.

Richert, M., D. Fisher, F. Piekniewski, E. M. Izhikevich, and T. L. Hylton. 2016. Fundamental principles of cortical computation: Unsupervised learning with prediction, compression and feedback. https://arxiv.org/abs/1608.06277.

Rilling, J. K., and L. J. Young. 2014. The biology of mammalian parenting and its effect on offspring social development. *Science* 345:771.

Rizzolatti, G., and L. Craighero. 2004. The mirror-neuron system. *Annual Review of Neuroscience* 27:169–192.

Rodrigues, S. M., J. E. LeDoux, and R. M. Sapolsky, 2009. The influence of stress hormones on fear circuitry. *Annual Review of Neuroscience* 32:289–313.

Sapolsky, R. M. 1996. Why stress is bad for your brain. *Science* 273:749–750.

Sapolsky, R. M. 2015. Stress and the brain: Individual variability and the inverted-U. *Nature Neurosciences* 18:1344–1346.

Schultz, W., P. Dayan, and P. R. Montague. 1997. A neural substrate of prediction and reward. *Science* 275: 1593–1599.

Schwartenbeck, P., T. Fitzgerald, R. J. Dolan, and K. Friston. 2013. Exploration, novelty, surprise, and free energy minimization. *Frontiers in Psychology* 4, Article 710.

Shadmehr, R., and Krakauer, J. W. 2008. A computational neuroanatomy for motor control. *Experimental Brain Research* 185:359–381.

Shadmehr, R., M. A. Smith, and J. W. Krakauer. 2010. Error correction, sensory prediction, and adaptation in motor control. *Annual Review of Neuroscience* 33:89–108.

Shouval, H. Z., S. S. Wang, and G. M. Wittenberg. 2010. Spike timing dependent plasticity: A consequence of more fundamental learning rules. *Frontiers in Computational Neuroscience* 4, Article 19.

Sigalas, M., M. Maniadakis, and P. Trahanias. 2017. Episodic memory formulation and its application in long-term HRI. In *Proceedings of the 2017 26th IEEE International Symposium on Robot and Human Interactive Communication (RO-MAN)*. New York: IEEE.

Simoncelli, E. P., and D. J. Heeger. 1998. A model of neuronal responses in visual area MT. *Vision Research* 38:743–761.

Solstad, T., C. N. Boccara, E. Kropff, M. B. Moser, and E. I. Moser. 2008. Representation of geometric borders in the entorhinal cortex. *Science* 322:1865–1868.

Solway, A., and M. M. Botvinick. 2012. Goal-directed decision making as probabilistic inference: A computational framework and potential neural correlates. *Psychological Review* 119:120–154.

Song, S., K. D. Miller, and L. F. Abbott. 2000. Competitive Hebbian learning through spike-timing-dependent synaptic plasticity. *Nature Neuroscience* 3:919–926.

Soulignac, M. 2011. Feasible and optimal path planning in strong current fields. *IEEE Transactions on Robotics* 27:89–98.

Sporns, O. 2010. *Networks of the Brain*. Cambridge, MA: MIT Press.

Sporns, O., and W. H. Alexander. 2002. Neuromodulation and plasticity in an autonomous robot. *Neural Networks* 15:761–774.

Srinivasan, M. V., M. Lehrer, W. H. Kirchner, and S. W. Zhang. 1991. Range perception through apparent image speed in freely flying honeybees. *Visual Neuroscience* 6:519–535.

Srinivasan, M. V., and S. W. Zhang. 1997. Visual control of honeybee flight. In *Orientation and Communication in Arthropods*, edited by Miriam Lehrer. EXS, V. 84, 95–113. Berlin: Springer.

Steels, L., and F. Kaplan. 2000. AIBO's first words: The social learning of language and meaning. *Evolution of Communication* 4:3–32.

Sugita, Y., and J. Tani. 2005. Learning semantic combinatoriality from the interaction between linguistic and behavioral processes. *Adaptive Behavior* 13:33–52.

Swanson, L. W. 2007. Quest for the basic plan of nervous system circuitry. *Brain Research Reviews* 55:356–372.

Tani, J. 2016. *Exploring Robotic Minds: Actions, Symbols, and Consciousness as Self-Organizing Dynamic Phenomena*. Oxford, UK: Oxford University Press.

Tapus, A., M. J. Mataric, and B. Scassellati. 2007. Socially assistive robotics—The grand challenges in helping humans through social interaction. *IEEE Robotics & Automation Magazine* 14:35–42.

Teyler, T.J., and P. DiScenna. 1986. The hippocampal memory indexing theory. *Behavioral Neuroscience* 100:147–154.

Thrun, S., W. Burgard, and D. Fox. 2005. *Probabilistic Robotics*. Cambridge, MA: MIT Press.

Tikhanoff, V., A. Cangelosi, and G. Metta. 2011. Integration of speech and action in humanoid robots: iCub simulation experiments. *IEEE Transactions on Autonomous Mental Development* 3:17–29.

Tolman, E. C. 1948. Cognitive maps in rats and men. *Psychological Review* 55:189–208.

Tomasello, M.,M. Carpenter, and U. Liszkowski. 2007. A new look at infant pointing. *Child Development* 78:705–722.

Tops, M., Russo, M. A. Boksem, and D. M. Tucker. 2009. Serotonin: Modulator of a drive to withdraw. *Brain and Cognition* 71:427–436.

Trappenberg, T. 2010. *Fundamentals of Computational Neuroscience*. Oxford, UK: Oxford University Press.

Trimmer, B., B. Vanderborght, Y. Menguc, M. Tolley, and J. Schultz. 2015. Soft robotics as an emerging academic field. *Soft Robotics* 2:131–134.

Tse, D., R. F. Langston, M. Kakeyama., I. Bethus, P. A. Spooner, E. R. Wood, and R. G. Morris. (2007). Schemas and memory consolidation. *Science* 316:76–82.

Tse, D., T. Takeuchi, M. Kakeyam, Y. Kajii Y., H. Okuno, C. Tohyama, H. Bito, and R. G. Morris, R.G. 2011. Schema-dependent gene activation and memory encoding in neocortex. *Science* 333:891–895.

Turing, A. M. 1950. Computing machinery and intelligence. *Mind: A Quarterly Review of Psychology and Philosophy* LIX.

Turrigiano, G. 2012. Homeostatic synaptic plasticity: local and global mechanisms for stabilizing neuronal function. *Cold Spring Harbor Perspectives in Biology* 4:a005736.

Turrigiano, G. G. 1999. Homeostatic plasticity in neuronal networks: The more things change, the more they stay the same. *Trends in Neurosciences* 22:221–227.

van Kesteren, M. T., D. J. Ruiter, G. Fernandez, and R. N. Henson. 2012. How schema and novelty augment memory formation. *Trends in Neurosciences* 35:211–219.

Vogeley, K., P. Bussfeld, A. Newen, S. Herrmann, F. Happe, P. Falkai, W. Maier, N. J. Shah, G. R. Fink, and K. Zilles. 2001. Mind reading: Neural mechanisms of theory of mind and self-perspective. *Neuroimage* 14:170–181.

Watts, D. J., and S. H. Strogatz. 1998. Collective dynamics of "small-world" networks. *Nature* 393:440–442.

Webb, B., and T. Scutt. 2000. A simple latency-dependent spiking-neuron model of cricket phonotaxis. Biologiocal Cybernetics 82:247–269.

Wilson, W. J. 2010. The Rescorla-Wagner model, simplified. Albion, MI: Albion College. https://campus.albion.edu/wjwilson/files/2012/03/RWSimplified.pdf.

Yamamoto, T., T. Nishino, H. Kajima, M. Ohta, and K. Ikeda. 2018. Human support robot (HSR). In *Proceedings of Siggraph '18: ACM Siggraph 2018 Emerging Technologies*. New York: Association for Computing Machinery.

Yamashita, Y., and J. Tani. 2008. Emergence of functional hierarchy in a multiple timescale neural network model: A humanoid robot experiment. *PLoS Computational Biology* 4:e1000220.

Yang, G.-Z., B. J. Nelson, R. R. Murphy, H. Choset, H. H. Christensen. S. Collins, P. Dario, K. Goldberg, K. Ikuta, N. Jacobstein, et al. 2020. Combating COVID-19—The role of robotics in managing public health and infectious diseases. *Science Robotics* 5:eabb5589.

Young, L. J., and Z. Wang. 2004. The neurobiology of pair bonding. *Nature Neuroscience* 7:1048–1054.

Yu, A. J., and P. Dayan. 2005. Uncertainty, neuromodulation, and attention. *Neuron* 46:681–692.

Zaldivar, A., Asher, D.E., and Krichmar, J. L. 2010. Simulation of how neuromodulation influences cooperative behavior. In *From Animals to Animats 11: Proceedings of the 11th International Conference on Simulation of Adaptive Behavior*, edited by S. Doncieux, J.-A. Meyer, A. Guillot, and J. Hallam, 649–660. Berlin: Springer.

Zender, H., O. M. Mozos, P. Jensfelt, G.J.M. Kruijff, and W. Burgard. 2008. Conceptual spatial representations for indoor mobile robots. *Robotics and Autonomous Systems* 56:493–502.

Zhang, H., and J. Jacobs. 2015. Traveling theta waves in the human hippocampus. *Journal of Neuroscience* 35:12477–12487.

Ziemke, T. 2003. What's that thing called embodiment? In *Proceedings of the Annual Meeting of the Cognitive Science Society*. Merced, CA: eScholarship Open Access Publications from the University of California.

Ziemke, T., D.-A. Jirenhed, and G. Hesslow. 2005. Internal simulation of perception: A minimal neuro-robotic model. *Neurocomputing* 68:85–104.

Zou, X., S. Kolouri, P. K. Pilly, and J. L. Krichmar. 2020. Neuromodulated attention and goal-driven perception in uncertain domains. *Neural Networks* 125:56–69.

Index

Intelligent Robotics and Autonomous Agents
Edited by Ronald C. Arkin

Billard, Aude, Sina Mirrazavi, and Nadia Figueroa, *Learning for Adaptive and Reactive Robot Control*

Dorigo, Marco, and Marco Colombetti, *Robot Shaping: An Experiment in Behavior Engineering*

Arkin, Ronald C., *Behavior-Based Robotics*

Stone, Peter, *Layered Learning in Multiagent Systems: A Winning Approach to Robotic Soccer*

Wooldridge, Michael, *Reasoning About Rational Agents*

Murphy, Robin R., *Introduction to AI Robotics*

Mason, Matthew T., *Mechanics of Robotic Manipulation*

Kraus, Sarit, *Strategic Negotiation in Multiagent Environments*

Nolfi, Stefano, and Dario Floreano, *Evolutionary Robotics: The Biology, Intelligence, and Technology of Self-Organizing Machines*

Siegwart, Roland, and Illah R. Nourbakhsh, *Introduction to Autonomous Mobile Robots*

Breazeal, Cynthia L., *Designing Sociable Robots*

Bekey, George A., *Autonomous Robots: From Biological Inspiration to Implementation and Control*

Choset, Howie, Kevin M. Lynch, Seth Hutchinson, George Kantor, Wolfram Burgard, Lydia E. Kavraki, and Sebastian Thrun, *Principles of Robot Motion: Theory, Algorithms, and Implementations*

Thrun, Sebastian, Wolfram Burgard, and Dieter Fox, *Probabilistic Robotics*

Mataric, Maja J., *The Robotics Primer*

Wellman, Michael P., Amy Greenwald, and Peter Stone, *Autonomous Bidding Agents: Strategies and Lessons from the Trading Agent Competition*

Floreano, Dario and Claudio Mattiussi, *Bio-Inspired Artificial Intelligence: Theories, Methods, and Technologies.*

Sterling, Leon S. and Kuldar Taveter, *The Art of Agent-Oriented Modeling*

Stoy, Kasper, David Brandt, and David J. Christensen, *An Introduction to Self-Reconfigurable Robots*

Lin, Patrick, Keith Abney, and George A. Bekey, editors, *Robot Ethics: The Ethical and Social Implications of Robotics*

Weiss, Gerhard, editor, *Multiagent Systems*, second edition

Vargas, Patricia A., Ezequiel A. Di Paolo, Inman Harvey, and Phil Husbands, editors, *The Horizons of Evolutionary Robotics*

Murphy, Robin R., *Disaster Robotics*

Cangelosi, Angelo and Matthew Schlesinger, *Developmental Robotics: From Babies to Robots*

Everett, H. R., *Unmanned Systems of World Wars I and II*

Sitti, Metin, *Mobile Microrobotics*

Murphy, Robin R., *Introduction to AI Robotics*, second edition

Grupen, Roderic A., *The Developmental Organization of Dexterous Robot Behavior*

Boissier, Olivier Rafael H. Bordini, Jomi F. Hübner, and Alessandro Ricci, *Multi-Agent Oriented Programming*

Hwu, Tiffany J. and Jeffrey L. Krichmar, *Neurorobotics: Connecting the Brain, Body, and Environment*